Foreword

The events of late 1973 in the Middle East, and the repercussions on the world energy economy, were of such profound significance that we are still unwilling or unable to face up to their implications.

The era of cheap and abundant energy supplies has ended. As a former US Secretary of Commerce has put it 'Popeye has run out of spinach'.

At the same time, geologists have shown that in any likely scenario of the future, world production of crude oil and natural gas will begin to decline before the end of the century. The countries that are fortunate enough to possess these resources will be at an immense advantage over those needing to import them.

Alternative sources of energy, and in particular coal and nuclear, can be exploited to replace liquid hydrocarbons, but not without creating new problems for mankind. Unresolved questions of safety remain for nuclear power plants, and the disposal of fissionable waste products is still a developing art. With regard to coal, it is doubtful whether men can be induced to work in deep mines in sufficient numbers to increase production significantly, while strip mining has grave environmental disadvantages even for a country the size of the USA.

These considerations are gradually leading governments in the direction of energy policies that act on demand as well as supply, and in the literature the number of references to conservation measures increases all the time. So far, however, it is tacitly assumed by the industrialised world that no very fundamental change in our way of life is necessary.

The papers given at the conference on 'Energy and humanity' will help to provide the materials for a judgment on whether continued rises in energy consumptions are either feasible or desirable. Obviously at some point in the future it would have to level off; the question is whether the critical point of time has been reached yet. Those who argue for a laisser-faire policy of unrestricted growth will have to develop complex strategies of resource substitution and technological elaboration, culminating in the millennium of the fusion reactor. The alternative is the simple decision, from a conceptual point of view, to stabilise demand by the middle eighties. Growth has been the god of the western world for so long that such a reversal of attitudes would be intellectually painful; but if humanity is put first instead of second, the reformation is already overdue.

Lord Avebury
April 1974

Preface

It is becoming increasingly more possible to foresee the consequences of our present highly technological way of life. The outcome of our failure to exercise restraint in any one of a number of important areas can now be predicted, and the effect that this will have on the future of mankind is a cause for concern.

The future of life as we know it on the Earth is subject to many interconnected constraints, such as in mineral resource depletion, food production, tolerable pollution levels, energy production and population growth. We are concerned here with our attitude towards these last two, and probably principally important, items — ENERGY and HUMANITY.

The world population is increasing at an exponential rate. The average annual growth rate is about 2%, which corresponds to a doubling period of approximately 35 years. This means that by the end of this century, almost as many people as already exist will have been added to the population of the Earth. Energy consumption per capita is also increasing at an exponential rate of approximately 2%, taking the global mean annual value. The outcome of the superposition of these two exponential growth rates is another even steeper rate of increase in the world total energy consumption, of 4% a year. The total world consumption of energy, then, if present trends continue, will double in the next 18 years; in the developed countries in recent years the doubling period has been much shorter. Gross national product, however, shows a rough correlation with energy consumption, and so the policy of 'Continued economic growth' demands that the above trends persist.

The global distribution of wealth (measured by standard of living or energy consumption) and population are not coincident, 80% of the world's energy being consumed by the 30% of the population living in the industrialised countries. The USA (with 6% of the world's population) consumes 35% of the world's energy; while the world average annual consumption is about 2 tons coal equivalent per capita, the figure for the USA is nearer 12 tce. As the world average annual rate of increase in energy consumption exceeds (but growing from a much lower level) that of the USA, the latter's share of the world's energy will fall to 25% by the year 2000. The differential will still be large, however, and the world average will still be less than half the US figure of today. In less developed countries, the per capita figure is almost constant, as population increase and energy consumption are matched, and so the gap widens between these and the industrialised nations.

To raise the whole world's average per capita consumption of energy to the same value as that in the USA today would involve a sixfold increase. Additionally the USA is expected to consume as much energy in the last 30 years of this century as it has done all-told to date.

Clearly it is not possible for the world's resources to sustain a standard of living for all of us equal to that enjoyed now (or expected to be enjoyed in the future) in the 'over-developed' countries. We are all aware that the Earth is finite, that resources are limited, and therefore we must see that, of course such growth cannot be sustained indefinitely.

There is no question but that we shall reach a state of equilibrium; no exponential increase can continue for ever. The choice is in the means by which we achieve it, rationally or by a succession of crises and disasters, and at what level. The choice is ours, now.

How do we tackle the problem of meeting the increasing demand for energy? Should we strive to fill 'the energy gap' at any cost to the environment, by ever increasing exploitation and devastation and with the inevitable accompaniment of the generation of more and more atmospheric pollution? Should we, on the other hand, change our values and attack the demand by striving for a reasoned, global approach to economic usage of energy and conservation of fuel, and also for better birth control. A stable population and a drastic reduction in wastage and extravagance must be achieved, together with the establishment of an integrated policy for utilisation of energy (both nationally and globally) for Energy and Humanity.

This book presents the problem as it exists now, with a view to what might be the situation at the turn of the century. An attempt has been made to cover the existing sources of energy along with their associated problems, to assess what might become available in the future, and to define and determine an approach to reasoned and realistic objectives. The papers represent the views of the authors, though the proposals are the concensus of opinion between genuinely concerned people of many different walks of life.

Much of the material has been drawn from an international conference on energy and humanity, held by the SSRS at Queen Mary College, London, in September 1972. It is hoped that it will be a source of further discussion.

London

R.J. Crookes

Contents

Part 1
Introduction

1   Introduction

Prof. M.W. Thring
Queen Mary   College, London, UK

1   The relation between industrial development and energy usage

The first industrial revolution has been entirely based on the replacement of human craft labour by machines, and these machines in general have used enormously much more energy to do the jobs. For example, instead of one horse pulling a small carriage with two people in it, we now have 100 horses or more. If a single person drives in a large American car, approximately 1000 times as much energy is being used as if the same journey were made on a bicycle. It is only because we have had virtually unlimited cheap coal since the last century, and now an equally unlimited usage of oil, that we have been able to expand our mechanisation in the last 100 years to an enormous extent. Fig. 1, taken from King Hubbert, shows the exploitation of fossil fuels in a historical perspective. The scale of the x-axis is one unit = 1000 years, and the Figure shows how, between 500 years before the present time and less than 1000 years after the present time, the whole of the world's fossil fuels, which took millions of years to lay down, will have been burnt up.

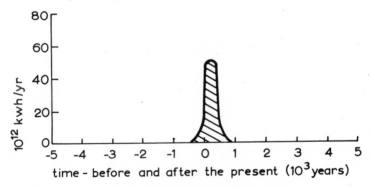

Fig. 1   Historical perspective of the exploitation of fossil fuels

The first question we have, therefore, to ask is whether less than 50 generations of human beings have the right to use up the whole of the world's fossil-fuel resources.

The second question relates to the more immediate problem of the running out of oil and natural gas, which is likely to occur early in the 21st century, so that a forward-looking policy to provide adequate facilities to our great-grandchildren is essential now. These resources will be exhausted if we continue to double their use for a few more decades.

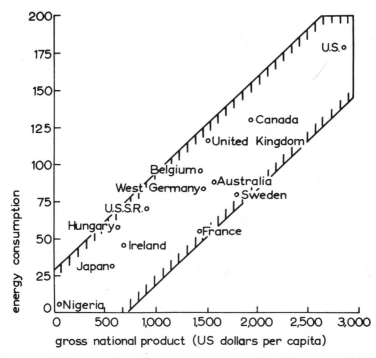

Fig. 2   Commercial energy use and gross national product show a reasonably close correlation

The third question is emphasised by Fig. 2, which shows the relation between the gross national product in various countries of the world in US dollars per capita, and the energy consumption. It will be seen that not only is there a close relation between these two, but also that the USA uses about seven times the world average (which is just under two tons of coal equivalent per head per year), while countries like Nigeria use one-tenth of the world average. Roughly speaking the developed countries are about one-third of the world's population with 80% of world consumption. The USA has been in a particularly favourable situation to develop an economy based on a highly extravagant use of fuel, because it has had its own plentiful resources of oil, natural gas and coal; although the oil and natural gas are now beginning to 'dry up' and the Pennsylvania oil and natural gas (which are first exploited) 'dried up' many years ago. Our third question is, therefore: what is the minimum energy consumption needed to give every human being in the world a fully adequate standard of living, and how far above this is the energy consumption in the developed countries?

At the present time, the total world energy consumption is doubling every 10 years, so that if it continues to do this it will be nearly eight times the present figure by the year 2000. The doubling period for the world mean consumption of energy per capita at present is longer at about 19 years, because the number of people is also rising. If we make the assumption that the required energy consumption per capita, to give a world citizen the full advantages of the first Industrial Revolution from the point of view of the necessities of life, is about equal to the present world average, then we could predict that the most optimistic solution would be that the rich countries would come down to the present world average and the poor countries would come up to it, and in this case the total world energy consumption by the year 2000 will only have risen owing to the increase in the number of people. However,

3

it is quite certain that the number of people will have nearly doubled by the year 2000, from the present figure $3 \cdot 7 \times 10^9$ to 7×10^9; thus the total world consumption even on the most optimistic estimate will be twice as much by the end of the century as it is now.

It is almost impossible to predict what will happen in the 21st century by the extrapolation of present trends, because, in the developed countries, the idea of a perpetually growing economy is still the official policy, and it is impossible to say at what point the existence of essential limitations to an excessively growing economy due to the limitation of the world's resources such as air, fresh water, land, fuel and other minerals will cause the growth to level off. Fig. 3 illustrates this point. As long as one is on the exponentially rising part of the curve (doubling every x years) and knows the ultimate world total, there are still several alternative futures possible according to whether

(1) the exponential curve is allowed to continue until physical exhaustion makes the resources harder and harder to obtain; in this case one obtains the classical symmetrical bell-shaped curve - practially a gaussian error curve - and the resource is exhausted as quickly as its useage accelerated.

(2) the world immediately adopts a policy of maximum conservation so that the curve levels off at the present rate of usage and then falls very slowly over several hundred years as the resource is used up. Such a conservation policy involves usage of the particular form of energy only for the purpose for which it is uniquely suitable (e.g. oil for transport) together with extreme economy of use for this purpose (e.g. public transport only) and the provision of identical facilities in all countries of the world.

(3) any alternative between these two extremes. I suggest, therefore, that a better way of finding out what we should do, andof answering our third question, is to study the conditions of a quasi equilibrium in the 21st century and to see the consequences of these. I call it a quasi equilibrium since a full equilibrium implies the use only of renewable resources such as solar energy or virtually unlimited resources such as Deuterium for fusion.

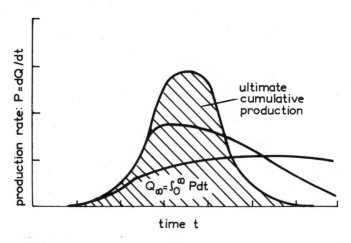

Fig. 3   Alternative growth patterns

If everyone in the 21st century (7000,000,000 people) is to have 2 tons of coal equivalent per head per year, it is equally certain that the world cannot afford to allow certain rich countries or rich individuals to use many times their share of energy. This means, essentially, that in the 21st century the gross discrepancies between the standards of living will have to be evened out. At first sight, this is a picture of a very drab world and one could say that one could only envisage it because the alternatives are world war or mass starvation.

However, if one makes a clear distinction between two important human factors, then this conclusion that the 21st century must be a time of gloomy drabness ceases to be valid. This distinction is between the standard of living on the one hand as expressed by GNP per capita or energy consumption, and on the other hand, quality of life. Quality of life is essentially a subjective thing; it is in each person the feeling of whether their life is worthwhile. A deterioration of the quality of life appears in the statistics of suicides and people opting out of life by drugs and alcohol, or setting up isolated communes, and also by the statistics of violence against individuals and of stress illnesses of all kinds.

Now the fundamental fallacy is to equate standard of living with quality of life. It is highly likely that the relation between them is a curve of the kind shown in Fig. 4. When the standard of living is very low, the quality of life is very low, because, in fact, one is starving and without the necessities to develop a life of any quality at all. Thus, for this region of the curve the two are almost linearly related. However, there is very little doubt that as the standard of living rises, an optimum point is passed, and one comes into the level represented by the idea of Kind Midas who could not touch anything without it turning into gold, and thus became the most unhappy person on earth. A society in which the only goals of life are the acquisition of more and more status symbols is, in fact, a society with a lower quality of life than one in which people are working hard to earn a bare living but doing work in which they find self-fulfilment. It also has a lower quality of life because of the fear of losing one's possessions or one's job, because of noise, accidents and pollution of all kinds, and because of the stress of being subconsciously aware that one is wasting the earth's resources while others are starving.

The peak of the curve must occur at the point where each individual has enough but only enough food, clothing, housing, privacy, social contacts, education, travel, opportunities for hobbies and leisure activities to develop his or her creative possibilities to the limit.

If the two tons of coal equivalent a year, which is the present world average, can be made to put everyone in the world at this optimum point without pollution, noise or accidents, then we can feel that we have left a fair legacy to our descendants in the 21st century and not a ravaged world full of our dangerous waste products and lacking the energy resources on which the rich countries have based their civilisation.

The fourth question which must be answered is how can we give everyone in the world for the next century, enough local and long-distance travel and transport to have a complete life within the available energy resources of the world, and without pollution, accidents, exhaustion of metal resources, noise, scrap iron left rusting about the countryside and much of the world's agricultural land covered with ugly concrete roads? Since, as has already been said, the most convenient fuel for road and air transport is liquid fuel consumed with air, we need to consider how to use a much greater fraction of the world resources of liquid fuel for this purpose and how to consume it more efficiently in terms of passenger miles per unit of heat and without the pollution and noise problems which are so unpleasant at the present time.

5

The first conclusion one would arrive at in this way is that, since liquid fuel is the only really conveniently available primary fuel for transport by air and sea and road and rail, liquid fuel should be conserved more and more for transport purposes; and since it is in short supply, we cannot afford to continue wasting it in large individual cars or other uneconomical transport systems such as supersonic flight. It is quite certain that a change of heart in this respect will not occur within 20 years from now, so that we must assume that the consumption of oil will continue to double until it is four times the present figure. However, this does not lead to a certainty that it will be completely used up, so we can base our study of the next century on the assumption that a reasonable amount of oil is still available purely for transport purposes, but that it has to be conserved very carefully basically by using only a public transport system so that the fuel consumption per passenger-mile or per ton-mile of goods is kept to an absolute minimum. Similarly, it will be 20 years before it is realised sufficiently by those who make decisions that natural gas is a rare and precious fuel and must be kept for those purposes for which it is specially suitable. Indeed, until recently natural gas in the Middle East has been flared off to waste and not used for any purpose at all, while its energy content was nearly as great as that of the oil with which it was associated.

The fifth question is whether we can make substantial use of nuclear fission energy, and later nuclear fusion energy, without the escape of radioactive materials, the danger of really serious catastrophes and without leaving highly radioactive materials behind, which must be looked after by our descendants for many generations?

The sixth question is whether it is possible to find a way of using the world's very substantial solid fuel resources without men going underground at all, and without the emission into the atmosphere of pollutants such as unburnt combustibles and sulphur dioxide?

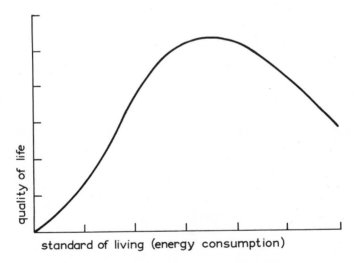

Fig. 4   Relation between quality of life and standard of living

Part 2
Energy supplies — resources and hazards

2.1 World energy resources

I. Th. Rosenqvist
Oslo University, Norway

1. Introduction

The sources of energy that are available for human consumption may be divided into two groups:

Group 1: regenerative energy

> The direct solar energy, which is basically of nuclear origin, and all those forms of energy which are directly derived from the solar energy, such as the food used by animals and men for their muscular work – this is mainly represented by living plants. Furthermore, we have fuel from wood, agricultural waste, peat etc. We have water power, the wind, tidal forces, and we may include geothermic energy, although this is not directly related to the nuclear processes in the Sun.

Group 2: nonregenerative sources

> (a) In this group, we have all kinds of fossil fuels which originally were derived from solar energy

> (b) Nuclear energy, which today is based on fission of uranium (and thorium), and in the future may be based on fusion of deuterium.

If we look on the earth as a space ship, it is of interest to clarify what we may call the 'ultimate energy resources'. The cost of energy is of less importance as long as there is sufficient available for the purpose we want to use it for, so that we may use engines instead of muscular power. It is said that 'no power is as expensive as no power'. As a unit of energy, it may be convenient to use 10^{10} ton coal equivalent (tce). This would correspond to the annual consumption of energy if all humanity should use the same amount as the countries of Western Europe. Some energy resources that today cannot be utilised economically, because cheaper types of energy are available, may be used in the future when the cheap sources have been exhausted. As the expense on energy constitutes a relatively low percentage of our total living expenditure, this will not necessarily lower our material living standard. On the contrary, any increase in the use of energy, nearly at any price, will permit us to increase the general production of material goods (which, in many communities, is called the living standard). There is, however, an absolute limit to the energy resources, dictated by the energy consumed in exploiting the energy resources. This has to be considerably less than the energy we gain. With this limitation, we may look at the known reserves of energy, and also at what is geologically probable. We may call this the 'ultimate resource'.

As any energy produced has to be consumed, it is interesting to examine the consumption pattern. In a technically highly developed country with a very high consumption of energy, like my own country, Norway, the total usage of energy is divided among the following groups:

Power-consuming industry (electrometallurgy etc.)	27%
Other industries	29%
Communication, excluding ships	7%
Domestic use and agriculture	25%
Other consumers and loss	12%
	100%

We do not know how energy will be used in the future, but it seems that power-consuming industries will take a relatively smaller part of the total.

Because of the progress in geology, geophysics and geochemistry during the last 20 years, we can today calculate with a great degree of certainty the amount of undiscovered reserves of energy. Consider a sufficiently large volume of the Earth's crust that is well investigated; it is highly probable that this will contain the same resources of energy and ores as an equally large volume of the Earth's crust of similar geological origin and the same geological history that has not been investigated.

In this way we can, on the basis of published data about known and proved reserves in well investigated areas, predict the ultimate amount of nonregenerative energy resources, coal, oil and uranium, even though we do not know exactly where to find the unknown and unproved reserves. General geological principles also permit us to predict the amount of reserves in those localities that have been partly examined, and where we have found resources of energy but do not know how much.

Because of taxation and for other reasons, most coal and oil producing companies tend to regard their proved reserves very conservatively; as for the fissionable raw materials, uranium and thorium, governments in many countries have forbidden the publication of data on the known reserves for political and military security reasons. Consequently, published proved reserves of these raw materials are very small, or not available. Even here we have different estimates, e.g. those by Euratom and IAEA, Vienna, ot how much uranium and thorium we may expect to find in the Western capitalist world, and also in the Eastern socialist countries and in the developing countries.

2. Transport of energy

Because the energy sources are unevenly distributed around the world, the cost of transport for different forms of energy is an important factor. The competitiveness of cost in various areas of the world is influenced by the cost of transport of the raw material. At the Pugwash SEAS symposium, held in London from the 12th to 17th April 1969, E.J. Nowacki from Poland gave a Table of the relative average cost of transport of energy in various forms (Table 1). This Table was presented at the World Power Conference in Moscow in 1968. The costs given are net costs, without tax. We see that, as long as we use large ships, sea transport is very cheap, and does not add significantly to the price of energy even over long distances. Transmission of electric power and transport of fuel by road are expensive, except for nuclear fuel. We can see a not so well known phenomenon, namely that it adds less to the cost of energy to transport oil from the Persian Gulf around Africa to North Europe than it adds to the cost of electricity to bring it across a relatively small country like Norway or Britain. The transport of electrical energy on high power line adds nearly as much to the cost of energy as transporting the equivalent oil in tank car over the same distance. As an increasing part of the energy is utilised in the form of electricity and not directly as heat,

Table 1

Relative average costs of energy transport by various transportation means

Form of energy and type of transport	Characteristics		Transportation costs cents/km/tce
Crude oil			
Tanker	80 000 DWT, 6000 km		0·04
Pipeline	24" - 30", 5-15 Mtce/year		0·10 — 0·15
Petroleum products			
Inland waterways	600 t,	500 km	0·40
Pipeline	1 Mt/a,	1000 km	0·50
Railroad	40 t capacity		
	car,	500 km	0·70
Road transport	30 t,	500 km	1·50
Hard coal			
Ocean collier	medium s.	6000 km	0·08
Inland waterways		500 km	0·28
Shuttle train		500 km	0·30
Pipeline		200 km	0·35
Railroad		300 km	0·75
Gaseous fuel			
Ocean tanker			
(liquefied gas)	medium s.	7500 km	0·20
Pipeline		1000 km	0·30
Electric power			
H.V. transmission	500-220 kV,	500 km	0·70 — 2·10
Nuclear fuel			
Water or railway			
transport		1000 km	0·01

economics will predicate that the fuel be brought to thermal power stations sited near the consumer, irrespective of whether the thermal energy is produced from fossil fuel or from nuclear stations. In the case of hydroelectric power, the consumer will try to locate his factory as near to the waterfall as possible.

The very large sources of hydroelectric power not yet utilised in South America, Africa and Asia may in the future be used for the high-consumption electrochemical and electrometallurgical industries. The difficulty is that such industries involve very high capital costs, and employ fewer men. This is just not what these countries might find most suitable or politically acceptable.

3.   Regenerative-energy sources

To illustrate what the Earth has of energy sources, let us, as mentioned, use the annual-energy unit of 10^{10} tce. This is a beautifully round figure in those countries that use the metric system (1 tce = 0·7 toe). The present world consumption is around 6×10^9 tce. In about ten years from now, we will probably be using the energy-unit figure. Using this as a yardstick, it is convenient to remember that the solar energy received by the Earth per year is $1·5 \times 10^{14}$ tce, and that we are losing the same amount to the universe by radiation; this is 15 000 times the energy unit we are dealing with. The gross total of photosynthesis on the Earth corresponds to a metabolism of $5 \cdot 5$ x 10^{11} tons of CO_2 per year, or a primary photosynthesis of ca. $1 \cdot 5$ x 10^{11} tce. This corresponds to 15 units of energy, or $0 \cdot 1\%$ of the solar energy. Nearly all this reduced carbon is again oxidised by the rotting of dead plants, or, in the living plants, by respiration during the night or other oxidation processes, because they act as food for animals and men. The production of plant material from the solar energy is about the only source of energy for living organisms, including man, and supplies the energy for all muscular work.

Through controlled burning of firewood, agricultural wastes etc., photosynthesis also contributes 300 Mtce to the human energy production, i.e. about 5% of the world consumption, mainly used in domestic household and in not very industrialised countries.

For the formation of new fossil sediments, it seems that, at the moment, we are only returning 5 x 10^7 ton of dry plant material. The present production of fossil carbon is consequently equal to a one-half percent of the assumed energy unit, calculated as coal equivalents per year. This return to the sediments is the equivalent of the net production of oxygen to the atmosphere by all the plants of the world. Those plants that are not brought into the fossil state do not produce any oxygen, because what they produce during their lifetime is consumed after their death.

The solar radiation is the source of another type of regenerative energy, i.e. the hydro-electric power. Because of the uneven evaporation and rainfall, it is possible to utilise this part of the energy received by the Earth. According to American estimates (US Department of the Interior, Geological Survey, Circular 483 1964), the exploitable hydroelectric-energy sources of the world are:

North and middle America	380 000 MW	
South America	690 000 MW	
Africa	1 065 000 MW	
Australia	205 000 MW	
Asia	1 300 000 MW	
Europe	290 000 MW	
Total:	3 960 000 MW	

Assuming the ratio of installed capacity to average useful capacity to be between $1 \cdot 5$ and 2 in those places where the hydroelectric power is already harnessed, and the mean gross capacity, when this figure is given, to be around $1 \cdot 3$ times the mean net capacity, we may recalculate the already builtup installations to kWh/year, and calculate what the unexploited rivers will give. Using an average recalculation factor of $1 \cdot 4$, we will see that, if all exploitable water power in the world could be exploited and optimally used, hydroelectric power would provide us with energy corresponding to 8 x 10^9 tce. (We assume that, on the producer's side, where fuel is used for electricity production, 1 kWh = 0·4 kg coal equivalent; and on the consumers's side, where the electric power is partly used to produce heat, 1 kWh = 0·25 kg ce). The exploitable amount of solar energy through water power consequently amounts to 0·005% of the solar energy received by the Earth.

Unregulated water power is a very cheap source of intermittent energy, but the regulated utilisation of hydroelectric power involves very high capital costs. Regulation is necessary, because electricity is only of value when it is produced in accordance with the demand, and this profile varies greatly through the year and through the day. No known source of energy can compete with well regulated hydroelectric power when it comes to following the variations between night and day consumption. For power-consuming industries, where the demand is nearly constant through the day and through the year, this property of the water power is less important, and only the price per kWh is of interest.

From the figures given, we see that, even if all hydroelectric power sources in the world were optimally utilised, they would not be able to deliver sufficient energy to satisfy future demand. In addition, there is the fact that they would deliver energy as electricity, which is uneconomical far away from the waterfalls, because of the transport costs. The present amount of the potential hydroelectric power exploited in the world is roughly 8%, in Europe more than 50%. Africa, Asia and South America have the greatest potential water power, but with the lowest percentage harnessed.

Other regenerative resources, beside the direct solar radiation, have from ancient times, been the wind, harnessed by windmills and sailing vessels; earth heat, in which recently we have got interested; and the tidal forces. In Iceland, Japan, Italy and a few other countries it is possible to extract useful energy from the geothermal gradient. However, mostly the thermal gradients are fairly low. The heat flow of the earth seldom exceeds 40 calories/cm^2 per year (i.e. enough to melt an ice layer ½ cm thick). Thus the exploitable parts of the earth heat are diminutive and only of local interest. To a certain extent, the same may be said about the tidal energy. The pressures are relatively low in most areas, and the amount of water that can be passed through turbines in power stations is limited. Tidal energy will in future mostly be used for the local production of electricity.

According to this review of regenerative-energy sources, it is obvious that humanity will either have to develop a technology that permits the use of the solar energy more economically, or rely to a great extent on exploitation of energy resources that are not regenerative. With the present technology, this means coal, oil, uranium and to a certain extent thorium. At the present time, it does not seem sensible to regard the content of deuterium in the oceans as an energy resource. The technology for utilising fusion reactors seems to the layman to lie so far in the future that it is to him beyond scientific possibilities to calculate when it will come, whereas the fission breeder reactor based on thorium may seem to be feasible within ten years.

4.   Fossil fuel

Classical fuel, coal, oil, gas etc., represents fossil solar energy. As stated in the previous chapter, gross photosynthesis today corresponds to roughly $1 \cdot 5 \times 10^{11}$ tce. We think that we have had the same degree of photosynthesis for at least 600 million years. It may have been going on for a much longer time in the oceans, and for a shorter period on land. This would correspond to 10^{20} tce of reduced carbon having passed through the photosynthetic cycles. Of this amount, 99·999% has been re-oxidised and returned to the atmosphere.

We are fairly sure of the amount of fossilised carbon the earth's crust contains. The first really good calculation was carried out by Prof. V.M. Goldschmidt in the 1930's. He was interested in geochemistry and distribution of the various elements on the earth and had already calculated and analysed the various types of rock. It was at that time generally accepted that all the oxygen in the atmosphere had originated from photosynthetic processes, and that the original atmosphere of the earth did not contain free oxygen. Furthermore, oxygen which had originally been produced from photo-synthesis had acted as an oxydant for inorganic compounds such as bivalent iron, bivalent manganese and sulphur. Since the amount of oxygen present in the atmosphere and the ocean corresponds to 232 g O_2/sq.cm, and the amount of oxygen that has been consumed in oxidising primary rocks and their content of iron, manganese and sulphur was estimated by Goldschmidt to be 256 g O_2 per sq.cm, he took it that the corresponding amount of carbon must have been fossilised. This corresponds to about 200 g of fossil carbon per sq.cm of the earth. In 1956, F.E. Wickman at Penn. State University calculated the amount of fossil carbon on the basis of the ratio between the stable isotopes C^{12} and C^{13} in diamonds, organic material and carbonates. He assumed that diamond represented the original isotopic ratio of stable carbon isotopes in the earth, and that the enrichment of the light isotope in biological carbon is an isotope effect. On this basis, he calculated that the earth's crust now contains 700 $\pm$ 200 g fossil carbon per sq.cm, which is in fairly good agreement with what Professor Goldschmidt arrived at thirty years earlier. Since the total surface of the earth is 5 x 10^8 sq.km, we can see that the Earth's crust contains 1-3 x 10^{15} tons of fossil carbon. This is 0·001% of the photosynthesis produced in the earth's history. We still have no better data than this.

Because the fossil carbon originates from organic life which once existed, it has a very particular distribution on the Earth, very different from that of other valuable elements, where only a very low percentage of the total in the ores can be utilised. The World Power Conference in Ottawa in 1936 estimated that 1·8% of the fossilised carbon that the earth contains might be utilised as fuel. The greater part of fossilised carbon is found as very low (average 0·4%) carbon schists and shales. Of this, 10^{13} - 10^{14} tons are probably present in sufficiently high concentrations to be considered as fuel when they are found. This shows that only under special geological circumstances will reduction products of plants accumulate in such concentrations that they will produce fossil energy resources.

At the 8th World Power Conference in Moscow it was concluded that the reserves of exploitable fossil fuel amount to between 1 and 2·5 x 10^{13} tce, but, given present technology, only 3·4 x 10^{12} tce is sure to be exploited economically. It is probable that this last figure is a relatively conservative estimate and even the first number may be a conservative estimate of what we call the 'ultimate reserves'. Whatever figure we accept as the economically exploitable reserves, we have reserves which may last for hundreds, or thousands of years with the assumed consumption of 10^{10} tce per year.

Of the economically exploitable reserves, 73% is hard coal, 15% is brown coal (lignite), and only 12% is exploitable shale oil, mineral oil and natural gas. The exploitable oil resources alone would hence only be able to cover the total energy consumption of humanity for 40-50 years, whereas black and brown coal reserves would last at least ten time as long. However, new findings in the Arctic, the North Sea and other offshore areas clearly demonstrate that, until recently, we have seriously underestimated the resources of petroleum.

As an illustration of the conservative estimates which have been made, it may be mentioned that, at the beginning of this century, the world oil production was 20 million tons per year, and the oil companies reported that world reserves would only last for 20 years. In the 1930s, oil production had passed 200 million tons, and the oil companies assumed that the reserves would only last for 30 years. Now the production has passed 2000 million tons, and the oil companies are evaluating their proved reserves to 5·5 x 10^{10} tons. Including the companies' known probable reserves, this will last for 40 years at the present rate of production.

It is true that those oil areas which were first exploited and are situated in the most industrialised areas, e.g. USA, have started to decrease in production, but other areas have taken over. Between 1962 and 1967 oil production was 9 x 10^9 tons, and the proved reserves increased by 19 x 10^9 tons. As examples, we may take the Ekofisk, Forties and Brent Structure oil fields, and Frigg gas field, in the North Sea. These fields were not even indicated at the time of the World Power Conference in 1968, they were first found in 1970; Ekofisk now ranks as No.19 among the world's 50 000 oil fields. This single field would have covered the world consumption in the year 1900 for a period of 50 years.

New paleogeological maps show that large submarine and lowland areas have been subjected to geological conditions under which we may expect the formation of petroleum. This is particularly the case for the areas surrounding the Arctic Ocean. The giant Tyumen oil field in West Siberia is said to be the largest oil resource of the Soviet Union.

On the basis of extrapolation from the data we have on young sediments in the Mexico Gulf and other sedimentation areas where the conditions are favourable for formation of oil, it seems that we have a geological production of protopetroleum of between 1 and 2 million barrels per km^3 of newly formed sediment. Consequently the oil-forming marine sediments will contain, in the primary state, 0·015% petroleum, whereas shale and schists have a content of fossil carbon of 0·4%. Thus the present rate of formation of protopetroleum in the world's sediments may be of the order of some few million tons a year, whereas consumption of oil is a few thousand million tons. Thus we are using more petroleum than is regenerated in nature.

In contrast with solid fossil fuel, petroleum, gases and liquids are probably never extracted from the original source rock. Normally they migrate to traps in a reservoir rock of sufficient permeability, which is covered by an impervious cap rock. Solid fuel may be preserved through geological folding and metamorphism without escaping, and increase in value from bituminous coal to anthracite, whereas the liquid and gaseous petroleum may easily be lost from the original source rock.

A small part of the petroleum generated during the earth's history has been accumulated in exploitable oil reservoirs. The necessary geological conditions are chiefly found in low-lying areas or on the continental shelves. We have to assume that, in future, the offshore sites will increase in importance.

During the last decade, production of oil and gas has grown rapidly, whereas less interest has been shown in solid fuel exploitation. Among the petroleum resources, shelf production has had an especially rapid development. According to one of the world's leading oil geologists, Lewis G. Weeks, the production in 1968 of offshore oil amounted to 16% of the world production, and it is assumed that ten years after that, i.e. 1978, offshore production will amount to 1/3 of total production. According to Weeks' estimate the ultimate oil reserves amount to nearly 10^{12} tons petroleum, i.e. 100 annual energy units in mineral oil alone. Recently, in Ocean Industry No.4, 1972, it was stated that Soviet scientists are of the opinion that half the world's oil will be found in the areas surrounding and below the Arctic Ocean. Geological maps make such an assumption not at all improbable.

Whereas liquid and gaseous petroleum will probably be increasingly produced from offshore fields, shale oil and tar sands will probably mainly be mined on dry land. Oil shales may, in some instances, be looked on as a petroleum source rock where the low permeability has prevented the migration of the produced petroleum to reservoirs. Consequently, the oil shales have kept their petroleum content even after geological movements, including foldings and faultings, but not metamorphism. In the more distant future, these sediments may gain in importance as sources of petroleum.

The organic compounds which are produced through photosynthesis and subsequent metabolic biosynthesis may be either relatively stable or unstable compounds. They may be divided into three groups: carbohydrates, proteins and fatty acids. The two first groups form the raw material in the production of solid coal and some gas, whereas fatty acids of the sapropel may give rise to formation of oil by catalytic processes on the surface of clay minerals (W.D. Johns and A. Shimoyama, 1972). We are only now starting to understand this process. It seems that 3-4% of the fossilised carbon will enter into the formation of petroleum, whereas 96% will remain as solid or form gas. There are two factors which affect enrichment and fuel concentration. The pure biogenic geological process permits direct fossilisation of carbon in high concentrations to form lignite or coal. Such conditions are present in the Everglades in Florida today, in the mangrove swamps at the Niger delta, and in other places where intensive biological activity takes place at sea level. At the same time, there is a sinking of ground due to the weight of the subfossil sediment formed from the organic material. The other form of concentration is the migration and concentration of petroleum into reservoir rocks.

There seems to be a higher probability for valuable concentration of petroleum than of coal. But, of the X . 10^{15} ton fossil carbon, 10^{13} ton may be useful coal and 10^{12} ton useful petroleum.

5. Raw material for fission energy

Whereas the special biological and fluid dynamical enrichment processes have lead to a very large part (1-5%) of the total amount of fossil carbon to be present as fuel, conditions by which the fissionable elements, uranium and thorium, are formed are fundamentally different. These elements behave in the same way as most other inorganic elements in the earth crust, i.e. they are relatively evenly distributed. Only a very small part of the elements is present in concentrations from which they may be extracted without having to use more energy than that present in the extracted fissionable material. Uranium is a typical lithofile element. It is strongly enriched in the earth's crust relative to the mantle and core. Uranium may be enriched by anatectic and hydrothermal processes in magmatic rocks and further enriched by sedimentary processes in placers, arkoses and organic and chemical sediments. Geochemical evaluations show that the earth's crust contains on average 3-4 ppm of uranium and 3 times more thorium. If we take the thickness of the earth's crust to be 10 English miles or 16 km with a specific gravity of 3, we see that a column of 1 cm^2 cross-section, 10 miles deep, will contain 10 g of uranium and 30 g of thorium. Since the earth's surface is 5 x 10^{18} cm^2, it contains 5 x 10^{13} tons of uranium, and 15 x 10^{13} tons thorium. This amount of uranium, used in conventional fission reactors, represents 5 x 10^{17} tce, and in breeder reactors 100 times more. In addition to the energy from uranium, we have the energy from thorium, which can be used in breeder reactors. However, only a very minute proportion of this uranium is present in such concentrations that it can be economically exploited. Because of military and political secrecy, we have no figures in the literature which correspond to the proved reserves of oil given by the oil companies. If we look at the official publications, e.g. the World Power Conference 1968, we find that the published proved reserves for the whole world only correspond to 51 000 tons in the lowest price bracket. This figure is obviously far too low. Euratom has calculated that the probable reserves of cheap uranium which are available to the western world amount of 456 000 tons of uranium at a price below 20 000 $ per ton U_3O_8. Other calculations (IAEA, 1970) give a total probable reservoir in this price group of 1·3 million tons, i.e. 0·000025% of the uranium in the earth crust. If we take the uranium deposits which may produce uranium oxide at a price of 40 000 $ per ton, we may possibly find that this amounts to 6 million tons U_3O_8. If the price is put up to 200 000 $ per ton, the reserves will amount to 30 million tons.

The uranium deposits which may produce uranium oxide in the cheapest price bracket fall in two groups. First those occurrences which contain so much uranium (more than 1000 ppm) that they can be exploited solely for uranium. These are chiefly precambrian fossil placers found in Canada and South Africa, further fossil arkoses in Western USA, hydrothermal enrichments associated with acid magmatic rocks, and some metasomatic deposits. The other source of cheap uranium are those deposits where uranium may be extracted as a biproduct in the mining of gold, phosphorous, copper, vanadium etc. In these cases the annual production of uranium will depend on the intensity of the production of main constituent. In the middle-and higher-price groups we have the bituminous and coal sediments e.g., Scandinavian alum shale, which in Sweden is calculated to represent reserves of approximately 1 million tons of uranium, whereas Norway, with the same types of rock, does not regard this as a uranium reserve at all. The reason for these different evaluations is that Norway considers the value of the uranium to be less than the value of the land which has to be destroyed if the shale

15

is mined. In these black shales the uranium is probably enriched by a combined biochemical and chemical process both during the sedimentation and immediately after the sedimentation during diagenesis.

We have reason to assume that a higher percentage of the total thorium will be found in detrital sediment deposits which may be economically mined. This is partly because the thorium-bearing minerals are more resistant to oxidation, which takes place during the formation of placer deposits. Thus after the earth's atmosphere became oxidising, the minerals have higher chance of surviving chemical weathering processes than have the uranium minerals. Once economical breeder reactors are produced, thorium may perhaps be one of the more important sources of nuclear energy.

Published proved reserves of thorium are, however, small, and amount to 1·6 million tons in the cheapest price bracket. We have reason to believe that the ratio between thorium and uranium will be greater than 3, in all price brackets.

Today we are consequently faced with the probability of about 1 million tons of uranium in the cheapest price bracket, and 6 million tons in the intermediate price bracket. These figures, although not proved, may be taken to have fairly high probability. Because of very intensive research since the latter half of the 1940s, partly motivated by the prestige involved in the possession of uranium by a nation, we believe that the situation that occurred in the oil sector, where the reserves were enormously under-estimated will not occur. In conventional reactors, 1 million tons of cheap uranium correspond to 10^{10} tce which is 1 annual energy unit. In breeder reactors, we have to subtract the integral energy consumed by regeneration and re-employment in a plutonium burner. Nevertheless, we may multiply the energy output by 50, so that the net energy production from U^{235} and U^{238} will amount to about 50 years with the world's present consumption. This is less energy than that of oil, and a very low percentage of what we can expect from total fossil fuel. Not even the uranium reserves in the middle price bracket will bring the energy content of uranium equal to that of classical fuel, even in breeder reactors.

From present world plans, by 1980 we will produce roughly 200 000 MW by atomic reactors. These reactors will chiefly give their energy as electricity, with a flat production day and night, summer and winter, and they will then supply about 5% of the world energy consumption today, or 0·03 energy units per year.

The energy produced by nuclear reactors will therefore only play a minor part in the world energy coverage. There will, however, be a spread of enriched uranium in many countries, and this, or the plutonium produced from it, is an important danger to world peace. Another factor is that the radioactive waste, if not taken well care of, may be a hazard to humanity. Perhaps the most important point against the spread of fission reactors is that they must be very large (more than 500 MW) before they can compete in any way with coal or oil. This makes this source of little value for industrially underdeveloped countries, which are poor because they lack capital and industry in the neighbourhood of the power stations. They cannot economically utilise the power produced in such concentration. For this reason, improved atomic power stations will mainly be an advantage for highly industrialised countries, and may increase their power of competition against the developing areas.

In the industrialised world, Great Britain seems to have a special faith in the nuclear energy. USSR and USA have much less. In 1971, USA produced $36·7 \times 10^9$ KWh on the basis of nuclear energy. This corresponds to half of the hydroelectric power that a small country like Norway with 4 million inhabitants produces. Nuclear energy represents 0·6% of the energy production in the United States, and may not even cover what is used in mining, extraction and enrichment of uranium. There was a myth, and many still believe in it, that atomic reactors will offer a nearly unlimited amount of cheap energy, with special advantages for the poor countries. There is little scientific foundation for such an opinion. The amount of raw material, the production price and the danger of environmental disturbances make fission energy a doubtful source of energy in the future.

6. Environmental problems in connection with utilisation of raw material for energy production

When high grade ores such as iron ore, coal or oil, are mined, the volume of rock taken out of the ground is larger than the volume of the waste which is produced during the operation. This means that the waste may relatively easily be brought back to the exhausted mines. Even though strip mining, e.g. in Tennessee, Kentucky in the United States, is destroying large areas, these areas may relatively easily be rehabilitated by backfilling and by the use of landscape architecture. The conditions are very different when ores of low tenor such as diamond and gold ores or uranium ores, in the middle and higher price brackets, are utilised. If tenor of an ore, such as the black shales, is of the order of 200 ppm uranium, this raw material will produce valueless waste products of a considerably larger volume than the rock from which it was extracted. Furthermore, such deposits are normally sediments of less than 50 m thickness, and can only be mined if they are close to the earth's surface, so that they can be mined in open quarries. They cannot be extracted from underground mines because of the extra cost. For this reason Norway does not look upon its uranium-bearing alum shales as uranium ore, because they form the underground in a densely populated area. Eventual exploitation of these alum shales will lay waste hundreds of square kilometers of fertile agricultural ground. All uranium ores in the world, which can give uranium in the middle and higher price brackets, involve an enormous destruction of landscape. Even if the producers do not take public opinion seriously, it is doubtful if many of the deposits of low-grade uranium ore will ever be mined, because governments do not want to have their landscape destroyed. In the highest price bracket we may find very large deposits with 40 - 50 ppm extractable uranium. A possible exploitation of 30 million tons U^{238} in this price bracket means that we have to blast and crush a volume of rock of 10 000 sq. kilometers down to a depth of 30 metres. In order to crush and chemically treat this volume of rock, enormous amounts of chemicals would be needed, and the volume of waste would be such that it is hardly feasable to produce a new artificial landscape. Thus, we can only take the uranium in the cheapest price and some in the middle price bracket as the world's reserve. For thorium the conditions are slightly better, but even here the reserves are small, relative to those of classical forms of energy.

Only if we limit fission energy to successful breeder reactors for part of the energy production in the highly industrialised countries will atomic reactors be of benefit to world energy production over a long period. This is because a limited emission of radioactive fission and spalation products such as Kr^{85} and H^3 might reduce the air and river pollution due to sulphur oxides which are produced by burning classical fuel. As radioactive and chemical pollution do not act additively, we may perhaps tolerate a small and controlled radioactive pollution.

To return to the oil and gas deposits, the direct hazard to the environment may, at moderate cost, be kept relatively small in the areas in which the fuel is produced. Petroleum is mainly extracted from boreholes, and the wells may be backfilled with pressurised water to decrease the consolidation settlement of the land due to the reduction of the porewater pressure and the increased effective stresses.

The atmospheric and river pollution due to the sulphur oxides produced by burning coal and heavy oil has a chapter of its own. The damage can be controlled, and they have to be treated on an international political basis. The damage to the environment produced by coal mining is dependent on whether they are produced from underground pits or by strip mining. The latter type of mining is the most harmful, however, in many cases the damage may be reduced.

7. Summary

The present author can only endorse the words of A. Parker, Chairman of the Consultative Panel on the Survey of Energy Resources at the 1968 World Power Conference:

'From the data summarised in the preceeding paragraphs, it is clear that the energy resources in the form of solid fuels, oil, natural gas, water power, and nuclear fission are sufficient to meet the growing needs of the world as a whole for a long period of time. Additional reserves will undoubtedly be discovered and there will be further development of methods of using the heat energy within the earth, and the income of energy available as direct solar radiation, tidal power, and wind power. The problem is that of economics, in selecting the sources of energy and the techniques to meet particular needs under the variety of conditions in different parts of the world. Relative costs change with developments and advances in science and technology. The subject is of great importance. General experience shows that with increases in the efficient use of energy there is a rise in the general standard of living of the people'.

8. References

1. BEAMISH-CROOKE, J.: 'World fuel supply and demand — Estimates for the years 1980 and 2000'. Background paper, SEAS symposium on economic aspects of energy production (with particular reference to nuclear power), London, April 1969

2. JOHNS, W.D., and SHIMOYAMA, A.: 'Clay minerals and petroleum-forming reactions during burial diagenesis'. Proceedings of the international clay conference, 1972, Vol. 11, pp. 233-242

3. NOVACI, P.J.: 'The technical and economical aspects of thermal power production'. SEAS symposium, London, April 1969

4. ROSENQVIST, I.Th.: 'Comparative economics of nuclear and conventional fuel'. First Pugwash symposium on control of the peaceful use of atomic energy with particular reference to nonproliferation, London, April 1968

5. WEEKS, G.: 'The Ocean Resources'. Off Shore, 20/6-68

6. WICKMAN, F.E.: 'The cycle of carbon and the stable carbon isotopes', Geochim. & Cosmochim. Acta, 1956, **9**, pp. 136-153

7. World Power Conference Survey of Energy Resources, 1968, London

8. 'The Geology of uranium'. (Chapman & Hall, translated from Russian, 1958)

9. 'Uranium exploration geology'. Proceedings of an IAEA panel, Vienna, April 1970

2.2.1(i)   Liquid fuels

C.A. Roast, C.Eng., M.I.Mech.E., M.I.E.E., F.Inst.F.
Esso Petroleum Co., Ltd., UK.

What is the worldwide oil supply and demand pattern likely to be by the year 2000? Prof. Thring recognises that there is no complete agreement among the various oil energy forecasters mainly because of the wide variety of influences. The general impression at the moment, however, is that, in the late part of 1990, the demand for petroleum products will be in excess of production.

In this highly competitive field of commercial enterprise, economic considerations generally take priority, but we cannot disregard the fact that, in the medium and long term, governmental measures with respect to crude-oil supplies on the one hand and environmental control on the other may have a marked effect on energy costs.

'Necessity is the mother of invention'; on this premise, I am confident that, in the year 2000, our children will enjoy higher comfort standards than those prevailing today.

Table 1

World Proven Crude Reserves *

Area	1960	1970
	$t \times 10^6$	$t \times 10^6$
Western Europe	236	508
Middle East	25100	47816
Africa	1110	9620
North America	4835	7400
Latin America	3430	3577
Far East and Australia	1500	1946
Sino-Soviet Area	4589	13698
TOTAL ($t \times 10^6$)	40800	84565

*   Institute of Petroleum

The published worldwide crude-oil reserves at the end of 1970 amounted to 620 billion barrels (84 500 x10^6 t). Worldwide consumption was around 18 billion barrels (2300 x 10^6 t). Provided that 1970 consumption rates were not exceeded in subsequent years, the then known reserves would satisfy the demand for at least another 30 years. There is, however, a cloud on the horizon, in so far as worldwide consumption of petroleum products has doubled each decade since 1950; if this pattern is maintained, the demand for the year 2000 would be around 150 billion barrels ($18\,500 \times 10^6$ t). If the demand rate of increase is maintained, we require a crude availability of 3 200 billion barrels ($440\,000 \times 10^6$ t) and the 1970 established reserves were 620 billion barrels. It is therefore implicit that we have to discover an average of 85 billion barrels ($11\,000 \times 10^6$ t a year for the next 30 years. This is equivalent to proving up, every year, reserves equal to a quarter of the vast total already found in the Middle East; it is indeed most doubtful that an average discovery rate of this magnitude can be maintained.

Table 2

Petroleum Products : World Consumption *

Area	1950	1960	1970
		$t \times 10^6$	
Western Europe	62	196	628
Middle East	13	35	51
Africa	6	13	43
North America	339	500	770
Latin America	42	80	135
Far East & Australia	26	76	316
Sino Soviet Area	45	144	346
TOTAL ($t \times 10^6$)	533	1,044	2,289

*   Institute of Petroleum

Various estimates of ultimate recoverable crude-oil reserves have been published in recent years; examples are as follows:

In 1965, Hendricks estimates were 2480 billion barrels. In the same year, Shell estimates were 1800 billion barrels. Ryman of Esso in 1967 estimated potential reserves to be 2090 billion barrels.

In 1970, Moody of Mobil estimated a reserve of 1800 billion barrels.

Warman, head of BP Geological Review Department, is of the opinion that the recoverable volume ranges between 1200 and 2000 billion barrels.

Assuming that the current trend of doubling consumption of petroleum products every decade is continued, a critical supply situation in the not too distant future appears to be inevitable.

What are the possibilities of making up this envisaged crude-oil shortage? Recent North Sea discoveries will make a small contribution, but it is apparent that we must direct our attention to the potential for synthetic crudes and Canada's Athabascan tar sands. There is now a better economic case for commercial development of the oil sands in Alberta than ever before. The recoverable reserves from these sands are estimated to exceed 300 billion barrels; this volume is somewhat significant when equated to the 1970 proven crude-oil reserves of 620 billion barrels. Technical difficulties, federal tax imposition and an internal surplus of Canada's crude supply has severely limited the amount of this synthetic-crude production. Although the pioneering venture at Fort McMurray continues to operate at a loss, there is complete confidence in a steady rate of progress towards economic viability. It has been postulated that the commercial development of Athabasca's synthetic crude must await a US oil-price level of $3·50 to $3·75 a barrel to become economically attractive. Today, US prices are in excess of $3·50 and Albertan crude is posted at more than $2·90. As conventional crude costs are rapidly increasing, exploration is becoming more expensive and producer-country governments are more active than ever before in extracting higher revenues from their indigenous wealth, the necessity to develop other forms of energy supply becomes urgent if current consumption trends continue.

What are the possibilities for synthetic-oil production from coal? Oil has been obtained from coal in the UK and Germany; some is currently being derived from coal in South Africa. Some economists have expressed the opinion that, in the USA, substantial volumes of oil will be produced from coal before shale is seriously developed.

Jamieson of BP is quite optimistic as to the future, and, although his forecasts (Fig. 1) may, to some authorities, appear too optimistic, his crystal-ball gazing may provide more substance than others have envisaged.

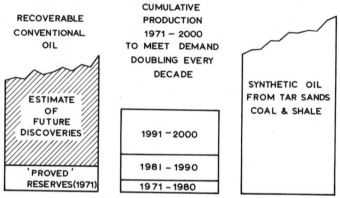

Fig. 1 Potential oil resources and demand (from W. Jamieson, BP Shield International)

Our immediate concern over the possible shortfall in crude-oil supplies by the year 2000 is based on the assumption that the demand for petroleum products will continue to double every decade. Is this a valid assumption? Although all the indications are that it is, there is justification for a study of the world consumption pattern as shown in Table 2 and the product band as given in Table 3.

Table 3

Breakdown of consumption product band (Excluding Sino-Soviet Area) *

Product	Percentage	
Gasoline	25·0	
Kerosenes & jet fuel	7·2	
Distillate fuel †	21·4	
Residual fuel	27·0	
Lubes and greases	1·0	
Other refined products	14·0	* Institute of Petroleum
Refinery consumption	4·4	† Includes Auto Diesel
TOTAL	100.0	

In Table 3, it will be noted that about 35% of the crude is converted into heating fuels used in industry and power generation. A valid question to ask is: what influence will nuclear energy have on residual fuel-oil utilisation for power generation? Some current estimates based on a capital-cost element of £174 per kW for a nuclear power station with £15 per kW as an initial fuel charge reveal a reasonably attractive proposition compared with a capital outlay of £45 per kW for a similar capacity oil-fuel power station. It would be foolish to assume that technical problems associated with nuclear power generation will not be resolved in the immediate future, and, in the circumstances, the next 20 years will show a rapid development in nuclear capacity. The UK, German and US governments are all assessing the future fuel requirements of their power-generation policies, and, instead of coal being phased out, its position in the prime energy league may well be maintained or even elevated. Any significant reduction in the percentage of the crude barrel processed into residual fuel may enable the refineries to produce more gasoline. While on the subject of gasoline, some thought can be given to the North American 1970 consumption of petroleum products (Table 2); it accounts for over 34% of the world total. Is this necessary? A breakdown of product usage would probably reveal that, of the 770 million tonnes, 60% or more was motor gasoline. A comparison between the miles per gallon consumed by the average American car with the average performance in Europe suggests that the price of petrol as a percentage of income in the USA is so low as to engender a complete disregard of thermal efficiency. The oil industry is confident that there are still many undiscovered potential oil fields, but it is expected that exploration costs will rise significantly. Now that the USA has legislated for large-volume imports of crude oil from North Africa and the Middle East, the long-term crude-supply situation will become critical in the event that world consumption rates continue to double every decade. It could be argued that a continuation of new discoveries of crude, such as those in Alaska and in the North Sea, will make a useful contribution. It may well be, that with the development of nuclear power, the current 27% of the barrel sold as residual fuel oil will be reduced to 15%, thus permitting, with, of course, additional refining processes, the production of more gasoline and aviation fuel.

By directing more of our natural-gas supplies to domestic heating systems, the barrel percentage of heating oil would fall in favour of more gasoline and aviation fuel. It is reasonable to anticipate rising costs for all petroleum products, and, in consequence, commercial development of the Athabascan oil sands will become economically attractive. In the USA and in Europe, there is a necessity to revitalise the coal-mining industry and thus permit it to make a larger contribution. The possibility of oil supplies running out by the year 2000 or even 2500 is remote. As to whether supplies at an economically attractive price will satisfy world demand is another question. The answer lies somewhere between the demand rate and the extent of nuclear-energy developments, the utilisation of solid fuel, the production of synthetic fuels from tar sands and coal, and, finally, the reorientation of the product band (Table 3).

References

Institute of Petroleum Statistics
BP Shield International
HR Warman: Petroleum Review, March 1971

2.2.1(ii) Solid fuels

David Broadbent
Director, Special Projects Department, UK National Coal Board, London, UK.

1   Introduction

I would like to take this opportunity to outline some of the problems of social responsibility in science. In so doing, I have no need to deviate from the title of 'Solid fuels' because, in the UK National Coal Board, we have a major basic industry employing some 300 000 people with a turnover of about £800 million, covering the complete spectrum of scientific, environmental and human problems. Furthermore, I believe that, as a nationalised industry, we should, and do, take the lead in reconciling the needs of the sciences, technology and the needs of people.

I believe that, in the future, we will see major changes in industry and in the energy industries in particular, and I would like to deal with the future energy scene and the 'energy gap' that will occur on present trends. My conclusions in this field follow very closely those of Mr. Roast, even though the approach is slightly different in some cases. Of even greater importance, in my opinion, will be the science-and-technology/ humanities gap, and I would like to offer some ideas on how these may be dealt with in the future.

I would then like to outline a possible future industrial and scientific complex where the material expectations of men are fulfilled, at the same time consistent with social responsibilities where the 'quality of life' will be a major criterion.

2   The need for coal and the 'energy gap'

2.1   The relationship between living standards and energy

Standards of living are achieved by drawing on the 'bank' of natural resources as though it was a deposit account. In the 1958-68 decade, world population increased 20%; chemical production 200%, minerals 57% and fuel 61%; i.e. a more than proportionate increase of resource consumption compared with population, leading to an overall increase in living standards. 10% of the world population consumed 90% of the world energy production, thus the rich becoming richer and the poor becoming poorer.

Per-capita energy consumption is a good guide to standards of living:

USA, 5299 kWh

West Europe, 1975 kWh

Africa 153 kWh

Industrialised countries 2683 kWh

Developing countries, 116 kWh

The problems of raising the gross national product of a typical African country up to UK standards are illustrated by the following back-of-the-envelope calculation. It would require 0·45 kW plant per capita at 50% load factor, and, with a conservative estimate of £100/kW installed, including transmission, it would require investment of £2,250 million for 50 million typical African people. If a factor of 10 is used for the cost of producing the equipment to consume the additional electricity, the secondary investment required amounts to £22 500 million, compared with an annual rate of investment in the UK of £9000 million for roughly 50 million people. This calculation shows the magnitude of the problem of raising the living standards of the underdeveloped nations and also that it is not possible to have sudden changes in living standards.

'Limits to growth.' — the Club of Rome Report, 1972 — estimates the lifetime of the following resources, based on present trends:

	Years
Aluminium	30
Copper	21
Iron	93
Lead	21
Mercury	13
Natural gas	22
Oil	20
Coal	111

You will see that the resources of the Earth are finite, and that we cannot continue with the present rate of depletion of these resources.

2.2 World energy and UK consumption

My case for the future energy situation has been made by Mr. Roast, and I am generally in agreement with his predictions. However. the following figures show a slightly different approach:

(a) Fig. 1 shows estimates of world energy consumption for both exponential growth and linear per-capita growth, together with estimated nuclear contribution.

(b) Fig. 2 shows the world projected demand for all fossil fuels.

(c) Fig. 3a shows total fossil-fuel 'energy gap' for the UK taking the best possible estimates of supply of nuclear energy and natural gas.

(d) Fig. 3b shows what nuclear power has actually achieved, compared with what theoretically should have been achieved to reach their targets for the year 2000.

(e) Even on the best possible estimates of the performance of nuclear power and natural gas, more fossil fuel will be burned in the year 2000 than at the present time. Any shortfall in performance will increase the size of the energy gap.

(f) All forms of energy supply will be required in the future.

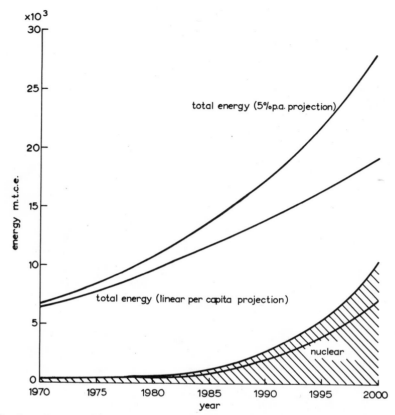

total energy (5% p.a. projection)

total energy (linear per capita projection)

nuclear

Fig. 1 Estimates of world energy consumption and nuclear contribution to 2000 AD

2.3 Coal or oil to fill the gap?

Fig. 4, prepared by H.R. Warman of BP, shows how world oil production has increased at an enormous rate since 1950, practically doubling every ten years or so. In the same period, however, reserves have not caught up with production rates, and Fig. 5 shows how reserves have fallen, from about 80 years in 1950, to about 35 years in 1970.

Forgetting the political problems associated with imported oil, the pressures on oil reserves will be enormous in the next decade or so, and the extraction of oil from the Earth's crust will become increasingly difficult and expensive. A particularly important factor will be the amount of oil that will be imported by the USA. J. McLean's recent lecture to the Harvard Business School in the USA referred to estimates that, by 1985, the USA will be importing oil at a rate equal to the entire output of the Middle East at current rates of production, and these predictions are generally in line with those of the US Academy of Sciences report, 'Resources and Man'. The report of John Irwin, US Under Secretary of State, to the OECD Ministerial Council in May this year was even more worrying. Mr. Irwin warned of a world oil shortage by 1980, and said that

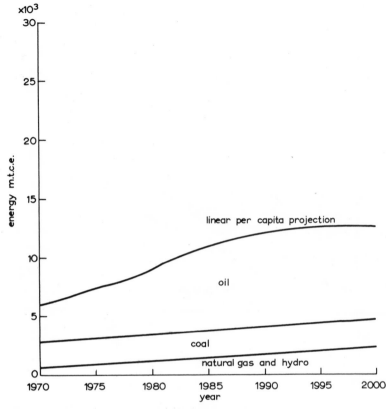

Fig. 2   Projected world demand for fossil fuels to 2000 AD

(i)   by 1980, US demand would rise to 24 million barrels per day and home (US) production (including 2 million barrels per day from Alaska) will only be 12 million barrels per day, i.e., by 1980, the USA will be importing 50% of its oil equal to 12 million barrels per day, roughly the amount that European members import today.

(ii)   The European members would import 20 million barrels per day by 1980, an increase of 10 million barrels per day. Oil from the North Sea would be not more than 3 million barrels per day.

Mr. Irwin concluded that it was imperative that the world's major consumers of oil and other forms of energy take joint and co-ordinated action to increase the availability of all other types of energy to lessen the dependence on Middle-Eastern oils.

There is, of course, oil in tar sands and oil shale, and work is being done in America to develop the technology for extraction of the oil. The key factor is the price of oil, and all indications are that it will increase substantially because of rundown of reserves coupled with the high cost of offshore working and transmission systems.

26

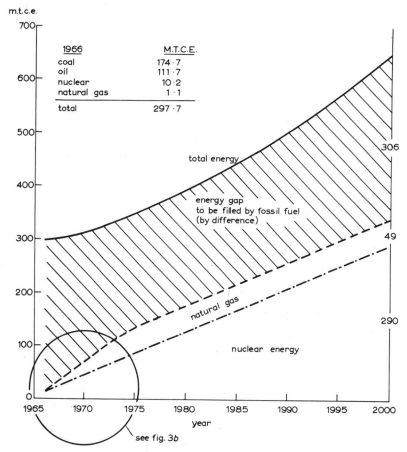

1966	M.T.C.E.
coal	174·7
oil	111·7
nuclear	10·2
natural gas	1·1
total	297·7

total energy

energy gap
to be filled by fossil fuel
(by difference)

natural gas

nuclear energy

306

49

290

see fig. 3b

Fig. 3a   Projections showing total fossil fuel energy gap up to 2000 AD (in UK)

Coal is the major source of fuel in terms of reserves. As mentioned previously, the Club of Rome Report gives 111 years' reserves for coal, compared with 20 years for oil. The US Academy of Sciences report gave 100 − 200 years for world reserves of coal. In the UK, we have known workable reserves of about 30 years, which probably represents only 3% of the reserves, so that, in effect, we have a potential of some years of coal reserves at the current rates of output.

In the USA there is an acute awareness of the fact that coal is the primary energy source and there is a massive programme for mining and for developing synthetic oil and pipeline gas from coal, as well as from oil shale and tar sands.

It is reasonable to postulate that, over the next decade, there will be a steady reduction in the difference between the cost of coal and the cost of oil, and, eventually, synthetic oil as a chemical feedstock will be produced from coal.

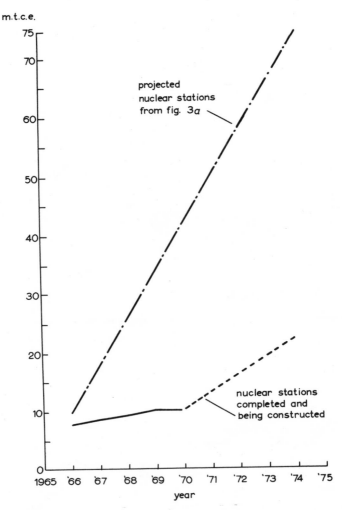

Fig. 3b   Nuclear stations — actual and projected

3   Coal industry of the future

The key factor for the future of coal will be the relative prices of coal and oil, so that the NCB is left with the old problem of reducing prices.

Without going too deeply into the commercial structure of the NCB, it is fairly clear from our accounts that about one-half of our costs are labour costs, so that, to reduce prices and increase productivity, men must be replaced by machines. There have been massive strides in productivity already in the mining industry, with mechanised coal-face operations now accounting for 92% of the production. Productivity has practically doubled from about 28 cwt/man-shift in 1960 to nearly 45 cwt/man-shift today.

bbls x 10^{11}

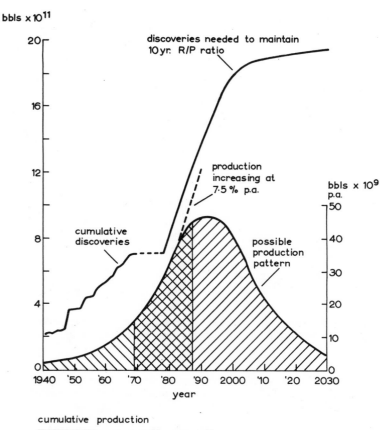

Fig. 4   Projected production and reserve requirement assuming ultimate recoverable reserves of 2×10^{12} bbls (from H.R. Warman, BP Ltd.)

Further increases in productivity must, therefore, be in improving mining techniques away from the face — taking the mining industry from a mechanised stage into the automated era.

Possibly, Prof. Thring's mining 'mole' could be developed, which, as he states, could possibly lead to mining processes 'being carried out without any human being going down the pit'.

There is, of course, an enormous human and industrial-relations problem involved in introducing these innovations. Imagine the miners being told that Prof. Thring's 'mole' had been developed to perfection and that we would not need any miners. The prospect of suggesting that we should be making some 300 000 men redundant is not conceivable, when one considers that the dockers struck over about 300 jobs and the postal workers are threatening strikes over about 20 000 jobs to be lost over the next few years.

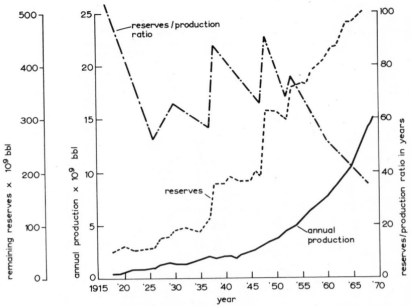

Fig. 5   World production, reserves and reserves/production ratio (reserves based on giant field discoveries backdated to year of discovery)

Clearly, no miner or docker or anybody wants to do work that is not needed, but, at the same time, he wants job security and an increased expectation of material goods in life, for himself and his family.

We have a situation in which there is unemployment and fears for job security, and pressure on industry to become more efficient and use less people. Unless there is a massive growth, it seems to me that the industrial strife that we see today is inevitable.

4   Possible solution

Today, we have a confrontation between science and technology, and the humanities — almost the title of the conference — and this is the major problem that must be solved before we can achieve what we all want, which is a reasonable quality of life.

We can divide any particular activity or undertaking into three major determining factors, namely:

(i)   technology

(ii)   management

(iii)   people and the human element

There must be a drastic rethinking at all levels of society, so that far more work goes into the third factor, the human element.

30

In dealing with the human element, both as an individual behavioural pattern and as a group behavioural pattern, we are at once beset by a host of unknown problems to which there are no card-indexed solutions or precedents to which we can turn for guidance. This is the area where we have to find a competence and a new dimension of understanding to allow us to make an impact on the people in our society.

We must also consider how the educational system has eliminated the difference between those who work and those who own the means of production, challenging the previous concept of master and man. The extent to which the leisure industries develop is a factor to consider, where such industries are creating work as an end in itself rather than a means to an end. All these matters are involved in the changing nature and character of work in modern industry.

There is one topic that does not require much debate, and that is that people today, in all walks of life, in society, and in industrial and other organisations, are sincerely challenging the what, why and wherefore of their role in the undertaking and even the very objects of that undertaking. This brings us back to the key factor in this debate, which is that we proceed along our present course at our peril. Any planning goals must, in the future, take account of the new thinking in the field of human attitudes and behaviour.

5 Future Industrial complex

I believe that, in the future, industry will have to take an increasing account of the needs of people.

I can envisage a shorter working week with guaranteed job-retraining schemes, and a floating pool of labour available and retrained in the skills needed for the industries of the future. The stigma of unemployment as we know it today will not exist, and job changing will be as commonplace and acceptable as was changing schools when we were children.

Environmental pollution will be strictly controlled, and industry will probably be in central complexes — some of which are being studied today in the USA. These centres could include metal reduction and processing, chemical industry and agricultural processes. Recycling of basic materials including waste will be a key feature in such developments. We, in the National Coal Board, are considering developments along the lines of a Coalplex system (see Appendix 7), where full resources of coal are used for heat, hydrocarbon materials (including pipeline gas), solid fuels, liquid fuels and chemical products including plastics. The ash associated with coal may be converted into useful materials such as cement, building materials and soil conditioners, and, with development, even such materials as aluminium, iron, nickel and copper could one day be 'byproducts'.

In the NCB, we are already carrying out environmental work on a considerable scale. We have established the use of pit shales as technically suitable filling material for motorways, such as the M62, and the Hovercraft terminal at Pegwell Bay. The clearing of tips removes eyesores, and releases derelict land for development. The 'total-cost' technique will be increasingly used — i.e. establishing the cost of land released, avoiding grants for future restoration — and it will enable a true identification of the costs of improving the environment. Opencast mining, although creating temporary dereliction, can be incorporated into large land-reclamation schemes. Yachting marinas or national parks can be created, as at Druridge Bay, and, in many cases, the environment would be better after the development.

6 Conclusion

My conclusion is that the future is in all our hands as scientists, industrialists, technologists, architects, engineers, planners and social workers, but, probably most important of all, as people. As members of a civilised and reasonable community, we have therefore the opportunity to plan a future in which economic growth itself will not be our prime aim, but improvement of the quality of life will be the goal to which we will all aspire. Recycling and better use of resources must be fully developed if we are to succeed in that aim.

To achieve that goal, we must also, in my view, put more emphasis on the needs of people, and do much more work on the social responsibilities of science, technology and industry. In fact, I would like to conclude by quoting the Duke of Edinburgh's recent speech made to planners and architects:

> 'For too long, we have become obsessed with the excitement of new
> technological developments, and forgotten that it is people and the
> living things of this Earth that are more important. We have rated computers
> above compassion, machines above mercy, and telecommunications above
> human relations.'

7 Appendix

Coalplex (Fig. 6)

There is little need to stress the possibility of an impending energy shortage towards the end of this century. In the USA in particular, it has been realised that, to reduce dependence on unstable foreign sources, there is no alternative to expanding the use of coal.

Causes similar to those giving rise to this potential upsurge in the use of coal in the USA will also affect Western Europe, including the UK. Europe already depends for about 60% of its energy on foreign imports, and this could rise to 80% by 1980. It is clear that a good case can be made for maintaining a significant level of coal production in Western Europe, at least as part of a policy of insurance.

One effect of the reduced worldwide availability of liquid and gaseous fuels over the next few decades will be an increase in cost of these fuels. In these circumstances, it is reasonable to postulate a reduction in the cost of coal, compared with oil, on a price-per-therm basis, and the use of coal as a substitute for oil will probably be commercially (as well as politically) viable in the UK within about a decade. This point of commercial viability will be reached in the USA (where the gap in price between crude oil and coal is already small) even earlier, if in fact it has not already been reached, given the successful technical outcome of some of the schemes for the better utilisation of coal now being considered.

With Coalplex, we are considering the use of coal as a source of energy, of coke and of gas and oil and of the products (manufactured carbons, acetylene and higher hydrocarbons, petrol and so on) now made by the petrochemical industry.

It is essential that the conversion of coal to these end products be carried out in plant that

(i)   operates at maximum efficiency, making sure that all the energy inherent in the coal is utilised

(ii)   is flexible, so that the energy and the various chemical products produced can be adjusted, as a percentage of the whole, to meet the required market pressures. It will also be necessary to get maximum utilisation of the major items of plant, by, for example, being able to regulate the production of storable products to take up the slack during those periods when the demand for electricity production is low. Such operation should make the best use of the labour force and allow good working conditions to be established

(iii)   minimises environmental pollution in all forms, both during the conversion processes themselves, and in the subsequent utilisation of the manufactured products

(iv)   can be integrated into the existing petrochemical industries network

(v)   is sited so that a market is available for the inorganic mineral matter produced, probably as a filler for building or civil-engineering work, though upgrading of part of the production, say, to lightweight aggregate, should be possible

It is with these requirements in mind that the idea of Coalplex installations has been evolved.

Coal has two main limitations as a feedstock compared with oil:

(a)   The hydrogen/carbon ratio (0·8:1 for coal as against 1·8:1 for oil) means that it is necessary either to take some carbon out or to add hydrogen at some stage in the conversion process

(b)   the fact that coal is a solid with a relatively high inherent ash content makes handling and processing more difficult.

Processes to manufacture chemicals, including petroleum, from coal, particularly that pioneered in Germany during the second world war, have been developed, though with limited commercial success.

More recent research, however, has opened up new avenues by which coal-to-oil or coal-to-gas conversion systems may be developed more efficiently than previously.

Two developments that it can be claimed were pioneered within the National Coal Board are very relevant to these new systems:

(a)   fluidised-bed combustion

(b)   solvent extraction, including extraction with gases.

Further developments in the UK, in this case by the Gas Council on methanation, have also made the technical feasibility of conversion systems more encouraging.

Fluidised-bed combustion is a method of burning (or partially burning if this is required) a fuel by injecting it into a cooled fluidised bed in which the carbon content is low. It has the following specific advantages:

(i)   Capital costs of plant are reduced, primarily because high heat transfer to surfaces immersed in the bed reduces the heat-transfer surface area required.

(ii)   Very high-ash materials or coals or chars with a variable ash content can be freely burned, so that coal preparation costs can be minimised and any low-grade products utilised. Hence the residues from coal preparation from the solvent-extraction plant can, for example, be conveniently and usefully incinerated. Running costs are also reduced, as fine grinding to p.f. sizes is not required.

33

(iii)   It is possible to eliminate virtually all gaseous pollution without special gas-cleaning plant. In particular, sulphur-dioxide emission can be prevented by trapping the sulphur in the bed by doping it with limestone. Nitrogen-oxide emission, especially under pressure, will be much lower, compared with other equivalent plant.

(iv)   It should be possible, since fouling and corrosion are reduced, to raise top steam temperatures above their present levels, hence improving the overall efficiency of the steam cycle and reducing generation costs.

(v)   Even more drastic improvements in overall generation efficiency and reduced plant size are possible by operating the combustor under pressure, as the exhaust gases can be cleaned enough to put through a gas turbine, making possible a combined cycle system.

Sufficient research and development work has been completed to demonstrate that this combustion system is technically feasible and that the advantages summarised above can be gained without venturing beyond the frontiers of already proved technology.

Solvent extraction is a means of obtaining the hydrocarbon content of the coal in a sufficiently pure form for further treatment by dissolving the coal in a suitable solvent, say anthracene oil, separating the undissolved coal and mineral matter, and then evaporating off the solvent. The resulting pitch-like material can be processed into, say, electrode coke or carbon fibres, or could undergo subsequent hydrogeneration to finish up as a synthetic crude suitable for further refining. In this latter case, it is possible in the process to increase the hydrogen content of part of the product, either by taking hydrogen from the remainder of the coal or from the solvent, so reducing the degree of subsequent hydrogenation required.

Studies are now in hand on the extraction of coal with low boiling solvents or gases under supercritical conditions of temperature and pressure, so that the extracted material is obtained as a gas. By this means, clean separation can be achieved between extract and undissolved residues, and the recovery of the solvent can be almost complete. It is also hoped that, as the process can be carried out at low enough temperatures to minimise thermal degradation (compared with liquid extraction), a more attractive range of products for subsequent processing can be produced.

Work by the Gas Council and the National Coal Board some years ago on the fixed-bed gasification of coal under pressure, provides an alternative approach to the coal-extract route, preferred in this concept. Other work by the Gas Council, including the catalytic synthesis of methane from carbon monoxide and hydrogen, for use as an alternative to hydrogenation, opens up new and improved routes from the products of coal extraction to the final production, either of enriched gas or crude oil.

The Coalplex concept has been conceived as a means by which advantage can be taken of these, and other, new developments by incorporating them in an integrated power and chemical complex feeding on coal.

In common with most projects, to develop more economical approaches to coal utilisation, specially those involving coal to oil or coal to chemicals, the Coalplex scheme is based on the idea of a large plant, directly taking all the output from a mine, or series of mines, and delivering crude oil and manufactured gas in significant quantities into a neighbouring refinery and gas pipeline terminal, respectively. By careful siting (say, in the UK, on the North East coast), it should be possible to maximise the economic benefits of low transport and the availability of onsite power, while the great flexibility of the industrial area as a whole, in feedstocks and in possible production patterns, should ensure the best economic return for all concerned.

The Coalplex scheme aims at meeting the requirements, as discussed earlier, of maximum efficiency and utilisation, flexibility, minimum pollution and the possibility of integrating into existing petrochemical industries.

In essence, the concept envisages:

(a) 'run-of-mine' coal being fed with the minimum of preparation straight from the coal face to the processing plant

(b) in the plant, the coal being passed down one of three main streams

(i) through a fluidised-bed combustor supplementing the waste fuels produced in, for example, the coal extract plant, to produce steam, heat and power, both for use in the complex and for marketing outside

(ii) through coke ovens to produce coke together with coke-oven gas and other byproducts. Alternatively (or in addition), it may be more attractive to add a formed coke plant. This would have advantages in increasing the range of coals that would be acceptable as a feedstock. The better flexibility of a formed coke plant and its suitability for intermittent operation will also help better integration. The distillation products would be fed into the chemical-processing streams from the coal-extraction plant [see (iii) below]; this would be a more economic use of these potentially valuable materials than the alternative of burning them for power generation

(iii) through a solvent extraction plant to produce a refined coal extract, all or part of which would then be hydrogenated or methanated to crude oil. Alternatively, or in addition, high Btu gas, or a range of carbon products, such as electrode coke and similar materials of high electrical resistivity, carbon artifacts and fibres or more sophisticated hydrocarbon products, could be produced.

These main streams, together with the subsequent stages of processing the coal extract, crude oil and coke-oven products (these stages being different depending on the types and quantities of final products required) will be fully integrated, and crosslinks, some of which have been indicated above, will be used to maximise the utilisation of the products at each stage.

Fig. 6 illustrates the broad basis of a typical design of plant.

A Coalplex installation will therefore produce:

(i) energy (mainly as electricity)

(ii) crude oil and gaseous products

(iii) coal extract for conversion to carbon products

(iv) coke for metallurgical purposes

(v) tars, creosotes and ammoniacal liquor, suitable [maybe associated with (iii)] for further upgrading

(vi) sulphur and sulphur compounds

(vii) fluidised-bed ash.

It is assumed that low-grade heat and steam will be utilised in the processes on the site, though it might be used for district heating.

35

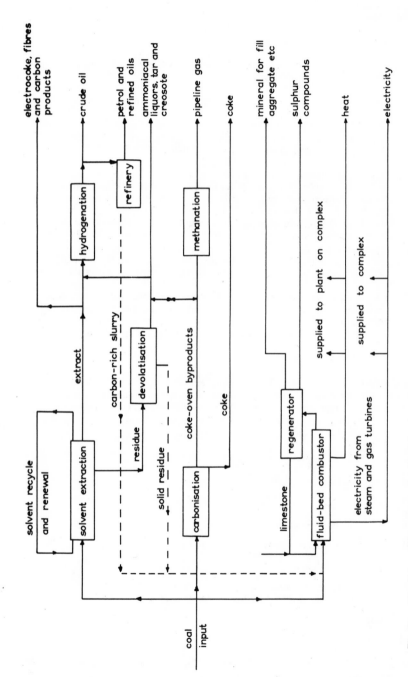

Fig. 6   Possible outline scheme for Coalplex

36

Markets for all, or most of the tars, creosotes, ammoniacal liquor and the sulphur and sulphur compounds should be found, though this must, of course, depend on the market condition at the time. The fluidised-bed ash should find a ready market, for manufacture into either light-weight aggregate or other building materials, or, failing this, as a filler material for civil-engineering work.

Other exciting extensions to the scheme outlined above are possible. For example, the availability of bulk hydrogen, either specifically manufactured in the complex or from some neighbouring source (nuclear power station), would allow further processes to be incorporated and the range of products that can be economically produced extended.

It is concluded that the Coalplex idea points the way by which coal can play the essential part that it must play in the overall energy pattern, both in the USA and in Western Europe, at the end of this century, and ensure the safe and healthy provision of clean, adequate quantities of energy at a reasonable cost with a minimum of environmental disturbance.

This paper outlines a few of the 'building bricks' from which a complex 'coal-refining' plant could be built up. It is hoped that it illustrates the possibilities that exist to use coal in a way by which maximum use will be made of all its potentialities, and by which the products formed from it can be made available to the consumer at a true cost that will be as competitive and cheap as possible.

What is required now is a concerted effort by fuel technologists and coal scientists, chemical engineers, planners and ecologists, economists and commercial people – in fact, everyone with an interest, direct or indirect, in such a complex as is envisaged in the Coalplex – to collaborate to produce a fully worked out, integrated plan.

2.2.2(i)   Environmental pollution

R.J. Crookes
Queen Mary College, London

There are harmful consequences of all power-plant operation based on energy conversion from nonrenewable resources, such as in the combustion of fossil fuels or in the fission of uranium. There are safety aspects and environmental effects to be taken into account even with conventional hydrocarbon-fuel burning plant, and these (although nothing like the magnitude of those connected with the use of nuclear fuels) require precautions to be taken.

In the 'production' of the fuels, safeguards must be incorporated at all stages. With coal, there is the danger associated with men working underground and the explosion risk, as well as the environmental effect of the waste products on land. With oil, there are dangers in offshore drilling, and pollution implications in transporting vast amounts by sea, and similarly, in the shipping of liquefied natural gas, there is a tremendous fire and explosion hazard potential.

Difficulties arise from the combustion process and the mechanisms involved, in actually converting the different forms of the fuel-energy source in the thermal plant, for here we invariably have to deal with extremely high temperatures, high pressures and often high-speed moving components, together with possibly noxious fumes, and the relevant safety measures must be taken.

The environmental impact of the burning of the fuels takes the form of noise, thermal and atmospheric pollution (which can in turn contribute to land and water pollution under the influence of wind and rain).

Different varieties of pollution are associated with different sources according to their scale, function, operating principle, type of fuel used and whether they are stationary or mobile.

In an analysis of environmental pollution, therefore, it is difficult to examine the contribution from any single source, such as combustion, in isolation from the many other sources. Nevertheless, the products and byproducts of the burning of hydrocarbon fuels can be seen to constitute a hazard to health and the environment.

In other chapters, we have been shown what the reserves of the world fossil fuel supplies are and at what rate they are being depleted. Burning is a very effective way of irreversibly converting valuable resources of energy into worthless, even harmful, waste products, and should therefore only be attempted under the most controlled conditions and at maximum overall thermal efficiency.

At first, man used fuel solely for heating and lighting, but subsequently, and most significantly, its usefulness in the production of power was discovered and exploited. Now, although conversion efficiencies are generally higher than ever before, the unprecedented use of energy means that atmospheric pollution is increasing at an alarming rate.

What is meant by the term Environmental pollution and when does a product constitute a pollutant?

Any event that, as a direct or indirect consequence of its occurrence, lowers the level of life quality, for any section of the community, or constitutes a harmful abuse of the environment, is pollution; any physical quantity or substance that, if in large enough concentration, causes physiological or psychological harm to man or damage to plants and animals or part of the natural ecology is a pollutant. Excessive noise, thermal burden or airbourne poisons fit this description.

1   Noise pollution

Noise is an unwanted sound transmitted by pressure waves through the air. It is emitted by combustion plant of all types in industry and in the domestic environment. Workers in industry should not be exposed to levels exceeding 90 dBA for any 8 h period, according to the UK Department of Environment Code of Practice (Reference 1), but levels higher than this are often encountered, and, in many industries, the employees do run a risk of suffering injury to their hearing. However, although noise produced by the heavy power plant of industry is usually specific to a particular locality, of more general annoyance to the community is the noise caused in transportation, which, by virtue of its mobility, penetrates almost everywhere. Road traffic, even where aircraft noise can be heard, is the principle offender, and can give rise to peak levels in excess of 90 dBA (Reference 2).

Like other forms of pollution, noise levels are rising yearly, and noise is now considered to be one of the main environmental pollutants. It affects communication by speech, causes stress and loss of sleep (possibly leading to nervous breakdown) and can induce deafness and lead to lower job performance levels. In addition to the loudness of the noise and the different reactions of people to it, there is also an effect arising from the variations within the noise climate (noises from different sources and of different character and duration) to which they are exposed.

It is very necessary that research be continued into establishing what realistic acceptable noise-pollution levels are and how these may be achieved for everyone in the community. Statutory legislation can help to implement them and put a limitation on present trends, for noise probably above all others is a pollutant endured mainly by those who contribute to it least. 'Peace and quiet' is part of the quality of life that should be available to everyone.

2   Thermal pollution

The useful energy attainable from a fuel in a thermal power plant, depends on the overall thermal efficiency, and the remainder, the 'waste heat', is transferred to the environment. Electricity-generating plants using the steam cycle, have an overall efficiency of about 30 - 40%, and so the bulk of the energy content takes the form of 'waste heat' and goes to the condenser cooling water (some is transferred to the atmosphere during transmission, and some leaves in the flue gases.) Of course the 30 - 40% actually available for use is eventually degraded to heat as well. The higher the efficiency, however, the less is the amount of fuel needed to obtain the same work output, and so the consequent thermal burden on the environment is lower. Solar energy is the only source of energy that does not add to this thermal burden.

2.1 Cooling water

Ecologists consider that temperature is the principal controller of life on Earth (Reference 3), and that thermal pollution of water constitutes a hazard to the balance of nature.

Using streams and rivers for direct cooling water on the scale envisaged in the USA over the next few decades is likely to become a threat to aquatic life there, particularly when combined with the pollution from sewage disposal. In 1968, 60 million million gallons of water were used in the USA for industrial cooling (Reference 3), and three-quarters of this by the electricity-generating industry. It is estimated that, by the year 2000, there will be 2 million megawatts of installed capacity in the USA, requiring one-third of the average fresh water run off for cooling water (in Summer, this could amount to the total runoff). A large power plant can raise the temperature of an entire stream by several degrees; so with several plants on one stream, the water would become uninhabitable.

Alternative forms of cooling are more expensive, such as the cooling towers more extensively used in the UK, and artificial lakes. Artificial lakes require large areas, and cooling towers, although increasing the cooling capacity of a river five fold (to about 10 000 MW installed capacity), emit tremendous volumes of evaporated water into the surrounding atmosphere. This can be in the region of 20 000 gal/min for every 1000 MW, which in Britain could amount to over ten tons per person per year (or one-eighth of our domestic consumption) with the local thermal gain going to the air. On cold days, this can result in fog and ice in the neighbourhood of the plant.

Clearly, something must be done in this respect, and the most logical approach is to use the waste heat to some advantage by substituting it for some other form of heating and giving consequent fuel economy in that application. What form this could take and how it could be implemented would depend on the particular plant, but suggestions have been made that use could be found in district heating, in irrigation and sea farming, or in desalination. The cities of Moscow and Warsaw are largely heated by passout steam from the power stations.

2.2 Atmosphere

The direct thermal contribution to the atmosphere, from man's combustion of fossil fuels, is a comparatively recent activity on the Earth, and is in fact the dissipation as heat, of the energy stored over millions of years of photosynthesis, from a small fraction of the Sun's incident radiation. Some ecologists consider that, should this contribution approach 1% of the Earth's current surface-radiation balance, then serious consequences would occur (Reference 4). At present, as a result of man's activities, large cities have climatic differences from their surrounding countryside, with higher mean temperatures and increased cloudiness and precipitation − they form urban heat islands. Another way in which man contributes to thermal pollution, leading to worldwide climatic and temperature changes, is by altering the atmospheric response to solar radiation, by increasing the carbon-dioxide and water-vapour content of the atmosphere, and so enhancing the atmospheric greenhouse effect. This means that, of the solar radiation incident on the Earth, part of that which reaches the Earth's surface and is re-radiated is trapped on its outward journey, being absorbed (at the longer wavelength) by the CO_2 and H_2O.

However, the situation is complicated by other aspects of atmospheric pollution, and by the fact that such a global temperature increase would also lead to higher evaporation. As well as contributing to the greenhouse effect, water vapour, in forming cloud cover (to which the condensation trails of high-flying jet aircraft may make a significant contribution), plays an important part in the control of the Earth's albedo. The albedo, or average reflectivity of the planet to incoming solar radiation, is also affected by the particulate-matter content of the atmosphere (its turbidity), and this is a function of both natural and combustion activity.

Fortuitously then, there may be a balance of these effects, but the situation is by no means well understood, and cannot be left indefinitely to chance . From the concept of thermal pollution, it seems that the problem of meeting man's demand for energy (and the likelihood of a forthcoming 'energy gap')must be approached from a conservationist standpoint.

3   Atmospheric pollution

All pollutants are increasing, many faster than the human population of the Earth, although, of course, they are closely related, and we have no way of knowing what might be the upper limit of the Earth's capacity to tolerate such a load without serious irreversible damage. In 1968, in the USA, 164 million metric tonnes of pollutants (Reference 5) were emitted into the atmosphere; half of this total came from automobiles. In the UK, the emissions are much lower, and the problem is not regarded as being so critical as in the USA, though in neither case are the pollutants evenly distributed, some areas suffering more than others.

What are the pollutants resulting from the combustion of fossil fuels?

When the combustion process takes place under ideal conditions, with exactly the correct mixture of clean fuel and oxidant, and progresses to completion, then the only products are carbon dioxide and water vapour. These are already present in the atmosphere and as such are not usually regarded as pollutants. Under such conditions, the combustion mixture is said to be in stoichiometric proportions. When excess air is present (to ensure complete combustion), the mixture is lean, and with insufficient air it is rich; and carbon monoxide, unburnt and partially oxidised hydrocarbons and soot are formed. Additionally, most fuels contain some impurities or additives, such as sulphur, lead (and other heavy metals), chlorine and incombustible inorganic matter, which lead to the formation of oxides of sulphur, hydrogen chloride and particulate matter. In lean combustion, the oxygen and nitrogen components of the air can react to produce oxides of nitrogen (Fig.1).

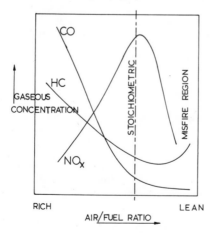

Fig. 1   Relation between mixture strength and pollutant concentration

41

3.1 Carbon dioxide and water vapour

The water vapour produced by combustion is only a small part of the tropospheric cycle (although it is highly localised and could be important) but the carbon dioxide generated is apparently altering the naturally controlled levels. 200 000 million tons (Reference 6) have been released so far by combustion (about 10% of the instantaneous atmospheric content), one-fifth of this in the last decade. The average concentration in the air has risen from 290 to 320 parts in 10^6 over the last century (Reference 7) and could reach 400 parts in 10^6 by the end of this century if current trends persist. The actual increase is only about one-third of that corresponding to all the CO_2 released by combustion, and it is apparent that most of this is being taken up by the oceans (some may be increasing land vegetation, though respiration and photosynthesis roughly balance). For how long this removal can persist is not certain, and a considerable influence on global temperatures may result in the future from the 'greenhouse effect' as already mentioned.

3.2 Carbon monoxide

About 200 million tons of carbon monoxide are emitted yearly into the Earth's atmosphere as a result of incomplete combustion. In the USA, one-half of the total mass of all pollutants is CO, and, of this, the motor car contributes 80% (compared with 50% in the UK). Steady-state combustors can readily be controlled to give far lower concentrations of carbon monoxide in their exhaust gases than in those obtained at present from the variable-power cyclic-operating internal-combustion engines. Apart from the pollution aspect, this is also a saving of fuel, as the CO has a considerable calorific value.

Concentration levels of 0·2 and 0·1 parts in 10^6 in the northern and southern hemispheres reflect the difference in automobile populations, but the absence of a rise in concentration level, comparable with mass of emissions, indicates that there is some sort of sink. The levels of CO are lower in the stratosphere, and so it might be that there is an oxidation reaction occurring above the tropopause (giving carbon dioxide) or alternatively, there might be a biological sink in the soil. The picture is by no means fully understood.

Carbon monoxide is a poisonous gas, which combines with haemoglobin to reduce the oxygen-carrying capacity of the blood. Exposure to a concentration of 50 parts in 10^6 for a period of 2 h can cause impairment of mental function, and such concentrations are reached in busy city streets in the UK, with peak levels reaching more than 100 parts in 10^6. Occupants of cars in traffic jams and underground car parks or tunnels are often exposed to much higher levels. For smokers, the situation is even more serious, because they already have CO in the blood.

3.3 Hydrocarbons

These again constitute a large energy wastage, and, in the USA, 60% of the hydrocarbon pollution comes from the internal-combustion engine of the motor vehicle (although here man's contribution is only about 15% of the total world emissions (Reference 8), of which most comes from rotting vegetable matter). In fact, the motor vehicles of the USA are responsible for the emission of as much hydrocarbon vapour as the motor vehicles of the UK consume (or about 5% of the US consumption). At high summer temperatures, hydrocarbons are also evaporated directly from the fuel tank.

Hydrocarbons resulting from incomplete combustion can range from unburnt fuel, cracked and partially oxidised hydrocarbons (aldehydes), to polycyclic aromatic hydrocarbons, which could well be carcinogenic. Unburnt hydrocarbons contribute to the unpleasant odour associated with the diesel vehicle, and play a significant role in the formation of the lachrymatory photochemical smogs of the Los Angeles basin.

Hydrocarbons have not been considered to be so great a problem in the UK, possibly owing to our more efficient, smaller units and, of course, our far lower per-capita consumption of fossil fuels, and lower summer temperatures.

3.4 Oxides of nitrogen

Oxides of nitrogen do not result from incomplete combustion; they relate closely to combustion efficiency, and are formed at temperatures above 1600°C. Nitric oxide (NO) is the major component of the total oxides (NO_x), and oxidises to the dioxide at lower temperatures in the presence of excess air. Just over one-third of the total US NO_x emission is produced by the motor vehicle, and over one-half in power generation and space heating. Continuous combustors, such as gas turbines and boilers, tend to produce high mass emissions of nitrogen oxides.

NO can form addition compounds in blood as does CO and NO_2 is capable of producing pulmonary oedema if inhaled. Oxides of nitrogen form nitric acid in rainwater, and, in London, over the last decade, concentrations have doubled. They are also a major component in the formation of the regional photochemical smogs that occur during inversions (when stagnant layers of stale air and combustion products are trapped under a layer of warm air preventing natural convection for several days) under the influence of strong sunlight.

They might also be capable of causing some climatic changes if released into the stratosphere (by supersonic transport) by reducing the ultraviolet radiation absorbed there by ozone, though this is hotly disputed.

3.5 Sulphur dioxide

Coal and heavy fuel oils used in the electricity-generating and other industries generally have some sulphur content, and nearly all the sulphur-dioxide pollution emitted into the atmosphere comes from this source. The emissions of sulphur-dioxide in the UK have increased from 5 to 6·3 million tons between 1952 and 1969, though urban ground-level concentrations have decreased by virtue of the 'high-stack' policy. In the US, emissions total 30 million tons. In the atmosphere, sulphur-dioxide is converted to sulphuric acid, forming acid rainfall, and, in the form, and in association with particulate matter, the life span can be considerably prolonged. In this way, it is possible for pollution generated in the UK to fall out in Scandinavia, reducing the pH of the water there and causing serious harm to aquatic life and to trees.

It is, however possible, if expensive, to remove the sulphur either from the fuel or from the flue gases, and this will be necessary as low sulphur fuels become scarcer.

The average concentration level in the UK (Reference 9) is about $106\mu g/m^3$ (0·04 parts in 10^6), but can reach $1000\mu g/m^3$ during inversions, and, at above $120\mu g/m^3$, respiratory disorders arise in the very young and old people (damage to plants occuring at much lower levels). During the 1952 London fog, which led to the 1956 Clean Air Act, 4000 people died, when sulphur-dioxide concentrations were as high as $4000\mu g/m^3$.

43

3.6 Particulates

The major sources of particulates are industry, power generation and space heating. Smoke, an airborne suspension of submicron soot particles, and grit have been reduced markedly from the industrial landscape in the UK since the inception of the Clean Air Act of 1956, which made it an offence to emit dark smoke, and the smokeless zones imposed on domestic areas have likewise produced a vast improvement. Similar acts have been introduced in the USA.

Vehicles are responsible for only 10% of man's contribution to the overall particulate pollution (which is an important factor in controlling the value of the Earth's albedo). Of the vehicular emissions, diesel smoke is probably the most familiar, consisting of fuel and oil fog, and soot (carbon) with associated polycyclic aromatic hydrocarbons. Probably more significant is the lead emitted from gasoline-powered automobile exhausts having been added (as tetra-ethyl-lead) to the fuel in quantities of between 2 and 4 g/gal to improve the performance. Lead is a cumulative poison, and the buildup of its concentration in the upper layers of the oceans and ice masses of the northern hemisphere is likely to be associated with internal-combustion-engine operation. Concentrations of $80\mu g/100$ g of lead in blood may produce poisoning in adults (Reference 10), with brain damage to children being possible at $40\mu g/100$ g, and yet many city children in the UK have blood levels that exceed these levels. Lead levels have doubled in Fleet Street over the last ten years, and many people may be unwittingly suffering low-dose effects, such as neural and behavioural effects like aggressiveness.

Once generated and emitted into the atmosphere, the pollutants are dispersed and diluted with air. Measurements can be made on samples taken for analysis from different places and at different times to follow the history and dilution of the pollutants, as they disperse from the vicinity of their source. Spreading from the street or factory over the town and valley and ultimately into the entire hemisphere (in the absence of physical or climatic barriers that would cause a buildup), the concentrations diminish progressively by orders of magnitude, and the mixing processes similarly increase from hours and days into months and longer. Different pollutants can interact in a synergistic way, and, although emitters claim that no harm is done at levels below known toxic limits, considerable discomfort can be caused at much lower levels.

It has been shown statistically that correlations can be made between air pollution (in a general sense) and health defects. Bronchitic mortality, lung and stomach cancer, infant mortality, the sickness of city bus drivers, and the difference in respiratory disorder between urban and rural postmen, have all been found to bear some relation to air quality. A saving of nearly 5% could be achieved in morbidity and mortality-related costs in the USA by a 50% decrease in air-pollution levels (Reference 11). The cost of air pollution in Britain amounts to £400 million, corresponding to £2 for every ton of oil equivalent burnt.

How can an improvement be achieved? In industry and electricity-generating plant, compliance with the clean-air acts, and use of 'smokeless fuels' and of the 99% effective electrostatic (and other) precipitators, have brought about an improvement in the visible particulate emissions. High stacks can only be considered to be a short-term solution to the SO_2 pollution, however, and cleaning the sulphur from the fuel or its oxides from the combustion gases must be considered as a high priority for future development. Improved combustion efficiencies and more extensive gas cleaning (washing) will become more necessary as pollution levels rise with increasing consumption of energy, and the reversal to coal combustion.

In the motor industry similarly, the US Environment Protection Agency (EPA) has demanded drastic reductions in emission levels from the manufacturers (in their automobile emission standards) to be achieved by 1976 (Table 1, from Reference 12).

44

Table 1

EPA Exhaust-emission standards

Component g.mile^{-1}	1973/74	1975	1976 *
HC	3·4	0·41	0·41
CO	39·0	3·4	3·4
NO$_x$	3·0	3·0	0·4

* Possibly subject to 1-year postponement

The currently popular afterburners and catalytic convertors, however, cannot be thought of as long-term solutions to meeting these levels, because they increase fuel consumption, and world resources of petroleum are limited. Research into different fuels for the IC engine and alternative engine designs (involving continuous combustion or the hybrid principle) will enable the lead to be excluded and pollutant-emission levels to be reduced, but generally with a reduction in 'performance'. This will inevitably have to be accepted, as will be the realisation that the freedom of the private vehicle will have to be restricted. No amount of improvement in combustion efficiencies will be able to nullify the effect of the ever growing numbers of automobiles and the increasing energy demands. Improved public-transport services for city travel and rail and waterway freight carriage, however, could go a long way in relieving the congestion, dirt and pollution in the city streets. If current trends (rates of increase in population and per-capita energy consumption) are not made to level off into some form of stable equilibrium, any antipollution measure will ultimately become ineffective. We must not only educate people on the facts of the environment (we have a responsibility to our neighbours and to our children, and the benefits of polluting are often removed in space and time from the costs), but we must also strive towards a stable population in the very near future.

The high cost of environmental protection does not get any less by delaying action, environmental degradation continues, and the difficulties of implementing the protection grow. A very real danger is that fuel shortages will be allowed to force a relaxation of antipollution measures, whereas the problems of fuel economy and pollution, can and should, be tackled as one.

4 References

1 UK Department of Environment: 'Code of Practice for reducing the exposure of employed persons to noise' (HMSO, 1972)

2 PRIEDE, T.:   'Noise in engineering and transportation and its effect on the community'. Conference on air pollution, Queen Mary College, London, April 1972

3 CLARK, J.R.:   'Thermal pollution and aquatic life', Sci. Amer., 1969, **220**, (3), p.19

4 EHRLICH, P.R. and EHRLICH, A.H.:   'Population, resources, environment' (W.H. Freeman & Co., San Francisco, 1971, 2nd edn.)

5 NEWELL, R.E.:   'The Global circulation of atmospheric pollutants', Sci. Amer., 1971, **224**, (1), p.32

6   STERN. A.C.:   'Air pollution — Vol. 3. Sources of air pollution and their control' (Academic Press, NY 1968, 2nd edn.)

7   BOLIN, B.:   'The Carbon Cycle', Sci. Amer. , 1970, **223**,(3) p.125

8   SINGER, S.F.:   'Human energy production as a process in the biosphere', Sci. Amer., 1970 **223**,(3), p.175

9   ROSE, S. and PEARCE, L.:   'Sulphur dioxide — a UK snapshot view' (Open University), New Scientist, 17th February, 1972, p.376

10   UK Department of the   'Pollution: nuisance or nemesis?' (HMSO, 1972)
     Environment:
11   LAVE, L.B. and SESKIN, E.P.:   'Air pollution and human health', Science, 1970, **169**, (3947), p.723.

12   Environmental Protection Agency: Federal Register, July 1971, **36**, (23), Pt.II, p.1265.

2.2.2(ii) Sulphur-dioxide emissions and other problems in future coal usage

Prof. R. Quack
Department of Energy Technology, University of Stuttgart, W. Germany.

The main part of my report covers SO_2 emission and how we can reduce it. But first I should integrate this special task with the whole field of problems that are connected with the future use of coal. We can see this field of problems under three headings:

(i)   Must we use coal or are there alternatives?

(ii)   If we must use coal, we should use it very efficiently.

(iii)   It is not sufficient to look only at the efficiency of the use of coal; we should use coal with the least harm to the environment.

Problems In Future Coal Usages

I *Must we use coal* ?   Alternatives ?
        Coal is used as heat source, it is perhaps
        more needed as a chemical substance.
    Other heat sources :
        Other fossil fuels or wood
        Municipal and industrial refuses
        Nuclear reactions
        Solar radiation
        Heat from the interior of the earth

II If we use coal, *we should use it*
        as efficiently as possible !
        Minimal loss in the pit (small seams !)
        Extraction of valuable hydrocarbons
        Complete burning with low heat loss
        Maximal energetic usage of possible
                        combustion temperature

III If we use coal efficiently, *we should do so*
        with minimal pollution of environment !
        Water pollution, Disposal of ash and slag,
        Noise and vibration , Radioactive radiation,
        Waste heat, *Air pollution*
        Dust and smoke   Disagreeable and poisonous
        (Sootless combustion, Gases : SO_2, NO_x, CO, Cl....
        dust precipitation) (Extraction before or after
                        Combustion)

Fig. 1

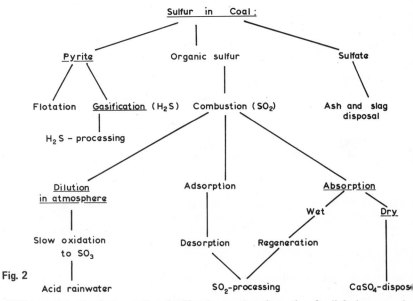

Fig. 2

With regard to the first question, I will only mention that other fossil fuels or wood are not the alternatives to coal as a heat source, but that those burnable refuses (that can not be made use of in other ways) are. In the future, perhaps the heat from the interior of our planet can replace coal as a heat source. I will not discuss all the methods of raising the efficiency of coal usage, but will only mention that the time of cheap oil and natural gas when it has not been economical to utilise all thermo-dynamic possibilities for better efficiency, will soon end. Therefore the combination of thermal power plants with the distribution of heat to housing and industry will again become a necessity. Let us now discuss the third heading; the minimising of environmental pollution in connection with coal usage. Under this heading we have several questions, but I will restrict the discussion to the problem of air pollution and, in particular, the poisonous-gas pollution problem of SO_2.

Coal, like some other fossil fuels, contains sulphur. Sulphur is an element that is needed in industry, mainly in the manufacture of sulphuric acid. Therefore, why do we not extract the sulphur out of the coal before burning it. The sulphur is in the coal mainly in two forms. One, the sulphate, is not of interest to us. It remains as calcium sulphate in the ash, and its solubility in water is so low that we can dispose of the ash and slag of coal furnaces without risk that the sulphate will be washed out and make the underground water dirty. Our problem is the sulphur, which is in the coal in the form of pyrite. This pyrite is distributed in the coal material in very small thin crystals. To get the pyrite out of the coal by mechanical means, we have to grind the coal to a very fine powder, so that each grain is either clean coal, clean pyrite or another inorganic matter. Then it is theoretically possible to separate the pyrite from the other ingredients by flotation, because the wettability of the pyrite is different from that of coal and other particles. But this method of obtaining pyrite, and hence sulphur, has some disadvantages. The coal with a diminished sulphur content, perhaps also with a lower ash content, is a moist slurry, which is difficult to handle and transport, and difficult to burn. Also it is milled to a greater fineness than is necessary for most of its uses. The disadvantages in obtaining pyrite by this method mean that the expense is greater than obtaining pyrite from pyrite mines. Of course, the pyrite out of pyrite mines is an ore, which is not inexhaustible. Besides the mines there are also other sources of sulphur, which make the price of sulphur as low as that of sulphur produced from coal. Such other sources are, for example, the mineral gas in Lacque in south western France with an H_2S content of about 15%, which must be

48

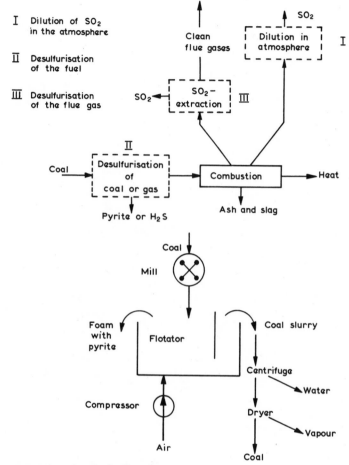

I   Dilution of SO_2 in the atmosphere

II Desulfurisation of the fuel

III Desulfurisation of the flue gas

Fig. 3   Extraction of pyrite by flotation

extracted before the gas can be distributed by pipe lines. Also, the steam which comes out of some boreholes in Italy and other volcanic places in the world contains more H_2S than is good for turbines and condensers. I think that we will get desulphurisation of fuel oil long before desulphurisation of coal becomes technically standard.

If we really have to desulphurise coal before burning it, I think that we will not do so by the separation of pyrite. We will first gasify all the coal, under reducing conditions, producing the sulphur as H_2S; the fly ash will then be removed; finally the H_2S will be extracted from the gas using a selective solution, for which process different solvents are available.

One disadvantage of this process, besides the high capital and running costs, is that the product of the gasifying of the coal is a hot gas, and the heat of the produced gas is lost, because the cleaning and washing of the gas is made at a low temperature. This heat loss can be reduced to some extent by a waste-heat boiler, but this must be of a special design so as to be inert to erosion by the dust in the gas.

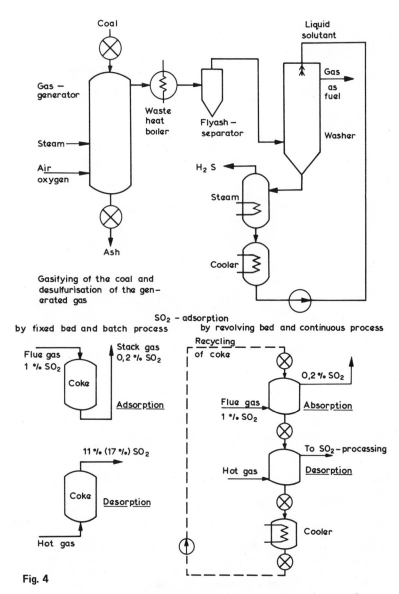

Fig. 4

The gasification of coal will be done at a pressure of about 20 bar, because the volume of the apparatus is thereby reduced and the gas is delivered at a pressure, which is favourable for its further transport and use.

The process of gasification, especially pressurised gasification, and of desulphurisation by extracting the H_2S, has been tested and could be used if this were necessary; but up to now it has been more economical to burn the coal with all its sulphur in it and hence produce SO_2 from the pyrite and the organic sulphur. Only a very small amount of the sulphur is oxidised to SO_3, and engineers are glad of this fact. Indeed, we like to forego the small heating value that could still be gained if we burnt all the sulphur not only to SO_2 but to SO_3, because we would have great difficulties with corrosion by SO_3 in the economiser and air preheater equipment. Therefore, power plant engineers are glad that the sulphur is contained in the flue gases of the furnaces, mainly as SO_2, and up to now it has been emitted with the flue gas through the stack or chimney to the atmosphere.

By the time all the sulphur has been oxidised to SO_3, each SO_3 molecule will have absorbed some H_2O molecules from the air, and, if it rains, the H_2SO_4 content will raise the acidity level, which will lower the pH value of the water of the rain droplets. The rain water in Scandinavia has a relatively low pH value, and the assumption that this is the result of the combustion of sulphur in fuel oil and coal in north western and central Europe is not unlikely. Knowing the pH value of rainwater, one must assume that rainwater is lacking in salt solutions, as very little H_2SO_4 in low concentration will lower the pH value greatly. This means, on the other hand, that the very low alkalinity of the ground, or the ingredients dissolved in the water that has reached the earth, can compensate for this acidity.

Why are we discussing SO_2 as a leading air pollutant? It is mainly because of the damage that it causes to special types of plant and to iron constructions, buildings and other works of art. Conifers, like pines and spruces, are very sensitive to SO_2, and the maximum concentration of 0·4 parts per million (p.p.m.) which is now allowed in Germany is mainly founded on experiments with these plants. Of course, it can be said that it is not necessary to plant only conifers in industrial regions; indeed botanists have already had some success in cultivating smokehard conifers. So we have to keep at the set restriction, even though we know that at 20 times that concentration there is still no risk to men and deers and other plants besides conifers.

Up until now we have solved the problem of keeping the SO_2 concentrations down by diluting the SO_2 contained in the flue gases in the atmosphere. But if the SO_2 emission of a single chimney is very high, because it is emitting the flue gases of several big furnaces or because the sulphur content of the coal burnt in those furnaces is especially high, the chimney must be very high to get enough dilution of the SO_2 in the air before the mixture reaches the Earth's surface. In one of our big coal-burning power plants, a chimney has now been built to a height of 304 m. There is, however, a problem in the dilution concept. To calculate the height of a chimney to get the desired dilution is only possible if we know the physical laws according to which the dilution takes place. The distribution of the stack gases in the atmosphere is dependent on many factors, of which the wind velocity and the atmospheric turbulence, among others, are variable.

In Germany now we have very pessimistic assumptions for these distribution laws, with the effect that much lower SO_2 concentrations are almost always measured in the air surrounding SO_2 emission sources than have been previously calculated. Sometimes it may happen that all the negative factors will coincide and cause the admissible limits of SO_2 concentration to be exceeded, in spite of the very high chimneys. Therefore the production of the SO_2 emitting industry must be reduced, or a fuel with a lower sulphur content should be used in those periods of over-emission, which has to be stored for such events.

When this industry was introduced to our country during the last century, there were already several laws to protect the neighbourhood against different dangers that were connected with industrial production. As well as these laws there were already some other laws against air pollution; but if the building of a factory has been allowed, perhaps with some restrictions in air pollution, it is difficult then to impose new restrictions in this respect. In Germany it was not until the last decade that new legislation made it possible to impose further demands on a factory with respect to admissible emissions.

What can an SO_2 emitter do now if he is forced to reduce the SO_2 emission of his factory. One possibility has already been discussed, to make his chimney taller, but this does not really diminish the emission, it only makes it more dilute.

The next possibility is to use a fuel with a lower sulphur content. This fuel is usually more expensive and not available in any great quantity. Two further possibilities remain: to desulphurise the fuel or to extract SO_2 from the flue gases. For this extraction, two main methods have been studied and tested: adsorption and absorption. In each case, the extraction of SO_2 from the flue gases is a difficult and expensive task, because the concentration of about 1% is already very low and to reduce it to an even lower level, extensive equipment is needed.

For adsorption, an adsorbent with a large surface area is needed and adsorption should take place at low temperatures. Charcoal or other very active adsorbents are, of course, very expensive, but coke, degassed at a low temperature of about $600^{\circ}C$, can be used. It has a relatively good adsorption capacity, but it is weak compared with high-temperature coke. Therefore it suffers from erosion if it is handled and moved. The adsorption capacity is limited, and, because it would be uneconomical to discard the used up coke, it is necessary to regenerate it. For that purpose the temperature of the coke must be raised because the adsorption capacity diminishes at a higher temperature. The higher temperature can be reached by two methods. One way is to use an external heat source by blowing a hot but inert gas through the coke bed. The other way is to blow a gas with some oxygen through the coke so that it will be heated by a partwise oxidation.

The result is regenerated coke and a gas with SO_2 in it. The result of the whole process is that the initial concentration of SO_2 in the flue gas was only about 1%, but in the regeneration gas the concentration is higher. All proposals to use an adsorption method to desulphurise combustion gases aim to obtain the SO_2 as liquid SO_2 or as fix sulphur or H_2SO_4. Therefore the concentration of SO_2 in the desorption gas should be as high as possible; 17% is often promised, but only 11% is usually the result. We cannot transport this gas over long distances, therefore the gas processing must be done close to the adsorption/desorption plant. An SO_2-processing plant is a real chemical factory and, hence, needs specially educated and trained personnel. However, we cannot afford to build such sulphur factories at each place where coal is burned. Therefore the adsorption process, in spite of being possible and technically tested, is never adopted in earnest.

Of course, it would be possible to transport not the desorption gas, but the SO_2-loaded coke. This would mean that beside each furnace we would have to install an adsorption chamber, but to install a common desorption plant in a big SO_2-processing chemical factory the coke would have to be transported to and from adsorption and desorption.

This idea has up to now not been proposed for the adsorption processes, because the amount of SO_2 adsorbable per weight of adsorbent is too low, and it is necessary to transport too much adsorbent between the plants. In this respect, absorption processes are better, because more SO_2 per weight of absorbent can be absorbed.

Absorption can be carried out by dry or wet processes. The difference is in the state of the absorbent leaving the plant, i.e. as a dry powder or as a liquid. The simplest absorption process is to blow the absorbent into the hot furnace gases and take it out of the flue gas in conventional dust precipitator equipment.

I have to mention this definition, because there are processes that inject the absorbent as a slurry and dry it by the heat of the gas, which is desulphurised, and the used absorbent leave the absorber as a sulphated dry powder; it is therefore called a dry process.

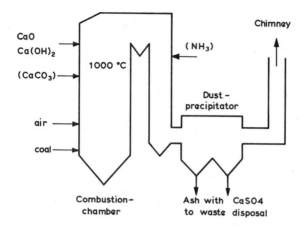

Dry Absorption Process

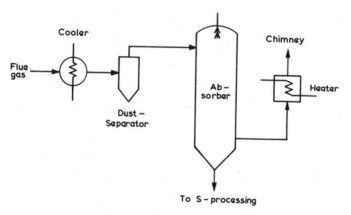

Wet Absorption Process

Fig. 5

Absorbents made up of several substances have been proposed, but according to our experiments the cheapest is the best, i.e. limestone (in the form of burnt lime CaO) as hydrated $Ca(OH)_2$ or as milled limestone $CaCO_3$. It can be blown in pneumatically as a dry powder or suspended in water as a liquid spray.

CaO and $Ca(OH)_2$ are best introduced at a temperature between 800 and 1000°C, because at higher temperatures the surface of the particles sinter and become less active. $CaCO_3$, which must be burnt by the heat of the flame, must be introduced at a temperature higher than 1000°C. The handicap of sintering will be compensated for by the fact that the decarbonisation effect produces a highly active surface.

This process produces a flue gas with greater dust concentration. Therefore the question is posed of whether bigger dust-precipitation equipment is needed to comply with the regulations about dust emission through the chimney, which are now, in Germany, 150 mg/Nm^3. If an electrostatic precipitator is used, and that is the most common equipment in modern plants, the sulphate-covered particles are precipitated with such great efficiency that generally no bigger precipitator is needed.

53

The dry absorption process, as a possibility to reduce SO_2 emission from coal fired furnaces, is mainly recommended in cases where, under normal weather conditions, the height of the chimney is sufficient to dilute the SO_2-loaded stack gases in the atmosphere, but where, under special weather conditions like no wind or inversion, the emission concentration limits perhaps may be exceeded.

There are absorption processes, most of them wet processes, that have a better desulphurising effect, and which are developed to obtain sulphur; but it is uneconomic to regenerate the absorbent at each furnace or power plant. It is therefore better to transport the sulphurised absorbent to a centralised S-processing plant, perhaps to a plant that is already producing sulphuric acid out of pyrite, because there we have the technological installations necessary for safe and economic processing.

One other proposal made in several countries, but only tested on a technical scale in Germany, is to burn the coal at a very low temperature in a fluidised bed, and absorb the sulphur at once with an absorbent that is also introduced in this fluidised bed. I have too much experience with the operation and possibilities of fluidisation to hope that this would be an optimal solution to our problem, but I am sure that by these experiments we will learn more about the physical and chemical laws which govern these technical processes.

To summarise, we must burn coal, then to do it efficiently, first, we can try to reduce the production of SO_2 by desulphurising the fuel. I recommend, not the extraction of pyrite, but the centralised gasification of coke. As long as we do not have such a centralised coal processing system and we continue to burn coal in normal coal furnaces, then I recommend, as a short-term possibility for times with dangerous weather conditions, that the SO_2 emission from the chimney be reduced by absorbing a part of the SO_2 with lime injected into the flue gases at a temperature of $800 - 1000^oC$, and by gathering the sulphated absorbents in the normal dust precipitator with which each modern plant is fitted.

I can only give a light recommendation to processes that adsorb or absorb SO_2 from the flue gas with the purpose of producing sulphur, liquid SO_2, sulphuric acid or ammonium sulphate. The only process that can be discussed in earnest is that where the loaded absorbent is transported to a centralised processing plant and, after regeneration, redistributed.

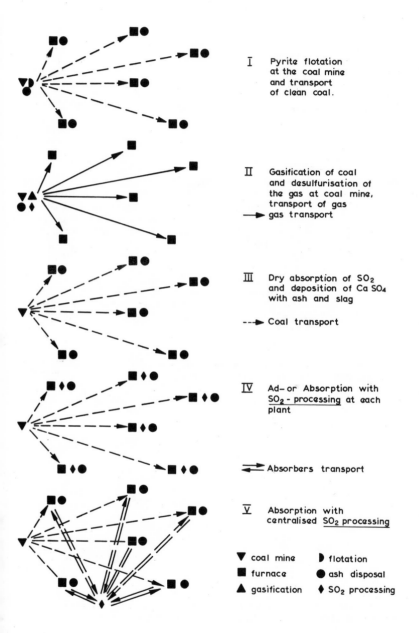

I   Pyrite flotation
    at the coal mine
    and transport
    of clean coal.

II Gasification of coal
    and desulfurisation of
    the gas at coal mine,
    transport of gas
    ⟶ gas transport

III Dry absorption of SO_2
    and deposition of $Ca SO_4$
    with ash and slag

    ---▶ Coal transport

IV Ad- or Absorption with
    SO_2 - processing at each
    plant

    ⇌ Absorbers transport

V   Absorption with
    centralised SO_2 processing

▼ coal mine   ◗ flotation
■ furnace   ● ash disposal
▲ gasification   ◆ SO_2 processing

Fig. 6

55

2.2.3 Nuclear energy and the future

Prof. P.D. Dunn
Professor of Engineering Science, University of Reading, UK.

1 Introduction

We will first of all summarise the present pattern of world energy consumption, and then move on to consider the total energy resources and the way in which energy requirements are likely to develop in the future. Table 1 gives the total annual energy consumption, and per-capita energy consumption for the world, the USA, the UK and several of the developing countries. The disparity between the developed and the developing countries is emphasised by these figures. The USA, with 6% of the world population, accounts for 33% of the current world energy budget.. If the current world population of around $3 \cdot 7 \times 10^9$ were to use energy at this rate, the annual consumption would amount to $4 \cdot 1 \times 10^{10}$ tonnes of coal equivalent (tce), or six times the current rate of consumption. Since, by the year 2000, the world population will have risen to around 7×10^9, the energy requirement would then be $7 \cdot 7 \times 10^{10}$ tce per year.

Table 1

Annual energy consumption 1970 (Reference 1)

Area	Energy consumption $\times 10^6$ tce	Per-capita consumption t/y
World	6843·25	1·889
USA	2282·32	11·144
UK	299·36	5·362
Ethiopia	0·79	0·032
Zambia	2·32	0·540
Mexico	59·17	1·205
Chile	11·81	1·208
India	102·67	0·191

Table 2 indicates how the energy totals are divided between fuel types. From this Table, we see the importance of liquid fuels, which account for about one half the fuel used in the UK and 43% of the fuel used in the USA.

Table 2

Breakdown of annual energy consumption by fuel type, 1970 (Reference 1)

Area	Total	Per capita	Solid fuels	Liquid fuels	Natural gas	Hydroelectric and nuclear
World	6843	1·889	2419 (35·3)	2850 (41·6)	1418 (20·7)	157 (2·3)
Africa	109	0·312	59 (54)	45 (41·3)	2 (2)	3 (3)
North America	2477	10·944	501 (20·2)	1025 (41·4)	897 (36·2)	54 (2·2)
South America	112	0·706	8 (7·1)	83 (74·1)	14 (12·5)	7 (6·3)
Asia (excluding Middle East)	540	0·480	184 (34·0)	325 (60·0)	16 (3·0)	16 (3·0)
Middle East	81	0·775	7 (8·6)	4·8 (59·2)	26 (32·1)	1 (1·2)
Oceania	77	4·03	36 (46·7)	37 (48·1)	2 (2·6)	3 (3·9)

Units: 10^6 tce and tce/capita

% figures shown in brackets

It is also of interest to look at the way in which fuel is used; this varies from country to country. From Table 3, we see that transport accounts for 30% in Western Europe and 24% in the USA. Transport is a major energy user in developed countries, and almost exclusively relies on liquid fuels because of their convenience, portability and high energy density. In general, fuels are not interchangeable for a particular application, and factors such as transport, storage and conversion to other energy forms must be taken into account. Our present pattern of energy usage has grown up around fossil fuels, and a change to nuclear fuels presents serious problems. There is no restriction in the size of the convertor for fossil fuels, which may be used in very small units for internal combustion engines or in very large units to raise steam for turbines. Fossil fuel can be easily and cheaply transported, and may also be stored safely and cheaply until required. The energy storage, both per unit mass and per unit volume, also compares favourably with other energy-storage mechanisms other than nuclear energy. This energy-storage feature of liquid fuels is a particularly valuable feature for transport applications. Fossil fuel is also convenient for both local heating applications and for central power stations. Having considered briefly our present pattern of energy usage, we should look at the total energy resources, to be able to assess the important energy sources of the future. Table 4 summarises the position. We see that, whereas liquid fuels are limited in extent, coal reserves are extensive. The renewable energy resources, such as hydroelectric power, wind power and tidal power, are unlikely to provide a major contribution to our long-term energy needs, although they may be of value in particular areas. An exception may be solar energy, and mention should be made of studies such as the Arthur D. Little proposal for large-scale energy generation by solar satellite. In this study, electricity is generated by solar cells, and is to be beamed down to Earth at microwave frequency and converted to normal mains frequency for distribution by the grid. The figures on nuclear energy require some explanation: the reserves of readily mineable uranium ore are small, compared with the oil reserves, but there is a virtually inexhaustable reserve of ore that could be used in breeder reactors. This aspect is discussed more fully in Section 2.3. Long term, we also have the possibility of fusion reactions again offering a virtually unlimited energy reserve.

Table 3

Use of energy in the USA, 1970 (Reference 2)

Use	%
Transport	24·6
Industrial	37·2
Residential and commercial	22·4
Conversion and transmission losses	15·8

Table 4

Total world major energy resources

Fuel		tce
Coal		$4·3 \times 10^{12}$ - $7·64 \times 10^{12}$
Crude oil	$1350 - 2100 \times 10^9$ bbl	$2·70 - 4·20 \times 10^{11}$
Natural gas liquids	$250 - 420 \times 10^9$ bbl	$0·50 - 8·4 \times 10^{10}$
Gas	$8000 - 12000 \times 10^{12}$ ft^3	$2·8 - 4·3 \times 10^{11}$
Tar sand	300×10^9 bbl	$6·1 \times 10^{10}$
Oil shales	190×10^9 bbl (10-100 gal/ton)	$3·8 \times 10^{10}$
Uranium (U_3O_8)	835 000 t reasonably assured)	
	740 000 t estimated additional)	at < 10
	1 575 000 t total)	$/lb.

2 Nuclear energy

There are three major types of nuclear reaction:

(i)   radioactive decay

(ii)   nuclear fission

(iii)   nuclear fusion.

Radioactive decay is not a significant source of energy, and, though it has some specialised applications, it will be ignored for the present application. We will first discuss nuclear fission, then nuclear fusion.

Nuclear fission

When a uranium-235 nucleus absorbs a slow neutron, the nucleus undergoes fission into two fragments of unequal size; in addition to the kinetic energy in the fission fragments, an average of two to three neutrons is also emitted. The major characteristics of this reaction are:

(i)   The energy released per reaction is very large, of the order of 200 MeV, compared with a few eV for a chemical reaction. One gram of uranium-235 completely converted by fission will give approximately 1 MW d of heat.

(ii)   Since fission is induced by capture of a neutron and the fission reaction releases a further 2-3 neutrons, a self-sustaining reaction is possible.

(iii) The energy release is in the form of random energy in the fission fragments, and hence appears as heat.

(iv)   A mass of fission fragments approximately equal to the mass of fuel used will be produced. Hence a 1000 MW power station will produce about 1 t of fission waste per year. The fission process does not always occur in the same way, and fission waste is made up of a large number of different elements, some highly radioactive and of long halflife. The problem of the handling and storage of this material is considerable.

Uranium-235 is called a 'fissile' material. It is the only naturally occurring isotope to undergo fission on absorbing a slow neutron. Natural uranium contains one part in 140 of uranium-235. The remainder of the material is the uranium-238 isotope. Uranium-238 absorbs slow neutrons, according to the reaction

$$_{92}U^{238} + n \rightarrow {}_{92}U^{239} \rightarrow {}_{93}Np^{239} + e \rightarrow {}_{94}Pu^{239} + 2e$$

Plutonium-239 is, like uranium-235, fissile and can be produced from uranium-238 by neutron bombardment. The latter is termed a 'fertile' material, and the process called breeding. Thorium-232 is also a fertile material, and is converted to the fissile uranium-233 by the reaction

$$_{90}Th^{232} + n \rightarrow {}_{90}Th^{233} \rightarrow {}_{91}Pa^{233} + e \rightarrow {}_{92}U^{233} + 2e$$

2.1 Types of fission reactor

Fission reactors are usually classified as fast or thermal. The former employ reactions with fast neutrons having velocities corresponding to the velocity at release in the fission reaction. Such reactors require the fuel to be almost entirely fissile material. The second type of reactor, on which most work has been carried out, is the thermal reactor, in which the emitted neutrons are slowed down by a moderator before re-entering the fuel material. A variety of materials are possible as moderators. In the UK, most effort has gone into graphite-moderated reactors, whereas, in the USA, the reactor programme has been concentrated on light-water moderation. An important advantage of thermal reactors over fast reactors is that the former do not require much, or in some cases any, fuel enrichment, and can operate on natural uranium. Both thermal and fast reactors can be used to breed; the former are more suitable for use with thorium-232 and the latter with uranium-238.

Nuclear fission reactors

We will, first of all, consider thermal reactors, and will divide them in terms of the method of moderation employed.

Graphite-moderated reactors

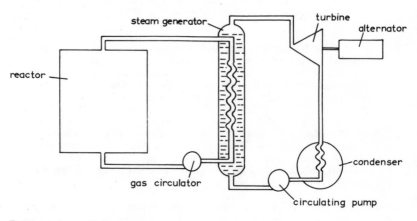

Fig. 1   Gas-cooled reactor

In the UK and France, most effort has been concentrated on graphite-moderated gas-cooled reactors (Fig. 1). In the earlier reactors, fission occurs in natural uranium rods sealed in finned magnesium cans placed in vertical channels in graphite blocks. In the later advanced gas-cooled reactors (AGR), slightly enriched uranium dioxide is used in stainless-steel cans. The graphite is arranged in the form of a right cylinder of height 10 m and diameter 13 m. The heat is extracted by CO_2 gas, which is passed through the channels. The complete core is contained by a concrete or steel pressure vessel. The hot gas is used to raise steam, which drives conventional turbogenerator equipment. The most advanced of these reactors can achieve a conversion efficiency of 40%, which is similar to that of a modern fossil fuelled station.

Light-water-moderated reactors

These reactors have been developed in the USA and installed in a number of countries. There are two types of water-moderated reactor, the boiling-water (BWR, Fig.2), and the pressurised-water (PWR, Fig. 3). In the former, water is boiled in the reactor and the resulting steam passed directly through the turbine, after which it is condensed and returned to the reactor. In this reactor, there is the possibility of contamination of the turbine by radioactive material escaping from the reactor. In the PWR, the reactor vessel is maintained at a pressure of $2250 \, lb/in^2$ and the temperature of the moderator maintained at $315^\circ C$; at this temperature and pressure, the water does not boil, and is passed through a heat exchanger and back to the reactor. The secondary side of the heat exchanger is maintained at a lower pressure of around $720 lb/in^2$, and is used to raise steam at a temperature of $260^\circ C$ for the turbine. Both types of water-moderated water-cooled reactor are smaller and cheaper to construct than a graphite-moderated gas-cooled reactor, but their conversion efficiency is much lower at a figure of around 32%.

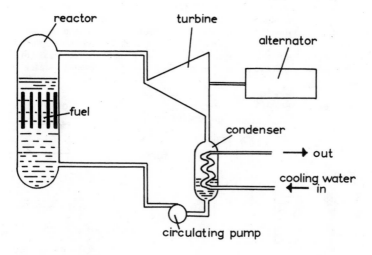

Fig. 2   Boiling-water reactor

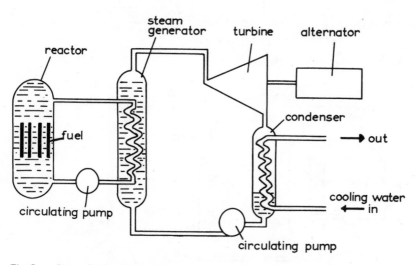

Fig. 3   Pressurised-water reactor

Heavy-water-moderated reactors

The Canadian programme has resulted in the development of the Candu, or heavy-water-moderated reactor; reactors of this type have also been built in India, Pakistan and Sweden. Moderation takes place in an unpressurised tank of heavy water, the fuel is contained in tubes, which pass through this tank and also provide a path for the coolant. In the Canadian reactor, this is light water at high pressure to prevent boiling. A heat exchanger is used to raise steam in the usual way (Fig. 4). The SGHW prototype in the UK is of similar design, but provides better steam conditions.

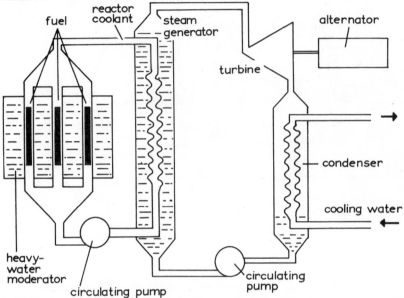

Fig. 4   Heavy-water reactor

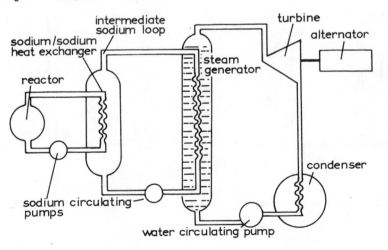

Fig. 5   Sodium-cooled fast reactor

Sodium-cooled fast-breeder reactor (Fig. 5)

The small, highly enriched core is cooled by sodium at atmospheric pressure; since the sodium coolant becomes radioactive, an intermediate loop is employed to separate it from the steam generator. Sodium is also used as the heat-transfer medium in the intermediate loop. The reactor both produces electrical power at an efficiency of around 40% and converts uranium-238 into the fissile plutonium-239. Table 5 lists the current position on these reactors.

Table 5   Liquid metal fast reactors (Reference 4)

Name	Country	Power MW(t)	MW(e)	Initial operation
		Operable		
BR-5	USSR	5		1959
Dounreay	UK	60	15	1959
EBR-11	USA	62.5	20	1963
Fermi	USA	200	67	1963
Rapsodie	France	40		1967
SEFOR	USA	20		1969
BR-60 (BOR)	USSR	60	12	1970
BN-350 *	USSR	1000	150	1972
		Under construction		
PFR	UK	600	250	1972
'Phenix	France	600	250	1973
FFTF	USA	400		1974
BN-600	USSR	1500	600	1976
		Planned		
KNK 2	West Germany	58	20	1973
JEFR	Japan	100		1974
PEC	Italy	140		1975
SNR	West Germany	730	300	1976
JPFR	Japan	750	300	1976
Demo 1	USA	750-1250	300-500	1978-1980
CFR	UK	3125	1250	1979
Demo 2	USA	750-1250	300-500	?
1000 MW(e)	France	2500	1000	1980
1000 MW(e)	West Germany	2500	1000	1982

* Construction of this dual electric power and desalting plant was completed in 1971

Other reactor systems

Over the last two decades, a great variety of reactor systems has been considered. Two of these are currently under study in the USA. The first is the gas-cooled fast reactor, in which the core is cooled by helium at a high pressure of $1250 \, lb/in^2$ and a temperature of $650^{\circ}C$ to give steam to the turbines at a temperature of $555^{\circ}C$ and, hence, a conversion efficiency of around 40%. One problem in this reactor is the provision of cooling in the event of loss of helium pressure. The second system of interest is the molten-salt breeder reactor (MSBR). This is a thermal reactor in which Li_7 - BeF_2 containing dissolved thorium-232 and uranium-233 circulates through a graphite matrix. The system can be connected directly to an online separation plant, and the extract temperature is sufficiently high to give a good conversion efficiency of nuclear energy to electricity.

2.2 Safety aspects of fission power: (Reference 5,6)

The major safety problems associated with nuclear power are

(i)   reactor accidents

(ii)   accidents during transportation of either fuel or fission products

(iii)   release of radioactivity during processing

(iv)   the storage of fission wastes.

The biological hazards and safe levels are discussed later.

Reactor accidents

Because of the new hazards posed by fission reactors, unusual care has gone into their design and construction, and their safety record is good. There have, however, been a few major incidents; for example, Chalk River in Canada in 1951, Windscale in 1957, and the Fermi fast breeder in Detroit in 1966. Precautions include the duplication of monitoring devices and an additional containment vessel of concrete or steel.

In most thermal reactors, the coolant is pressurised. In the case of the early Magnox reactors and the pressurised water reactors, the containment vessel is of welded steel, but the later Magnox and the AGR reactors use reinforced-concrete vessels. In the Candu and SGHW reactors, the moderator is not pressurised, and the coolant flows through an array of parallel high-pressure tubes. In these reactors, it is assumed that the maximum credible accident will be the sudden loss of pressurisation, in which case it is arranged that the reactor will be immediately shut down. Fission heating will then cease. However, considerable heat will continue to be released in the core for some time, owing to the decay of fission products. Initially, this heat release will amount to around 6% of full power for a gas-cooled reactor and as much as 10% for a water-cooled reactor. Hence emergency cooling arrangements are necessary. In gas-cooled reactors (Reference 8) air must be excluded from the hot graphite core, and emergency carbon-dioxide circulators are provided. The loss of pressurisation is particularly serious with the water reactors, owing to their high power density. Emergency cooling arrangements are installed but there is currently concern on the effectiveness of this emergency cooling (References 9 and 10). The sodium-cooled fast reactors pose particular problems, since, though they operate at around atmospheric pressure, they contain highly enriched fuel and have a compact core and, hence, a very high power density. The sodium coolant is highly combustible, and bubble formation raises the reactivity of the core. In the event of a coolant failure, the reactor is shut down, but the residual heat from radioactive decay in the core will amount to about 6% of the full power rating, hence core meltout is a serious possibility. The uranium-238 plutonium-239 fast-breeder requires particular care to avoid release of the highly dangerous plutonium. In this respect, the molten salt uranium-233 breeder is much safer, since uranium-233 is only weakly alpha-radioactive.

Transportation

There is a possibility of the release of radioactivity in the event of an accident occurring during the transport of spent fuel elements to the processing plants and also during the transport of the processed waste to the disposal sites. In addition, there is always the possibility of a dangerous release following a highjacking attempt.

Processing plant

There will probably always be some release of radioactive material during processing, but such plant will be operated under strict controls and should not pose a major problem. Plutonium-239 and uranium-235 are separated out and retained.

Disposal of fission waste

The fission process can occur in a variety of ways. Two of the most abundant pairs of fission fragments are strontium-90 of halflife 28 years with xenon-133 of halflife 5 days, and yttrium-91 of halflife 57 days with caesium-137 of halflife 30 years. It is found that the most active materials have a short halflife and decay quickly, whereas the isotopes of very long halflife are only weekly active. The two most troublesome isotopes that occur abundantly and have halflives of around 30 years are strontium-90 and caesium-137. These 'high-level' wastes amount to about 1% by volume of the total fission products, but contain approximately 90% of the total radioactivity. They produce a considerable quantity of heat, 200 W/kg after one year falling to 10 W/kg after three years. It is reported that the USA has accumulated 80 million gallons of these wastes over the past 30 years. Table 6 summarises the nuclear electric power existing in 1970. This total is expected to have increased by a factor of 100 by the year 2000.

Table 6

Nuclear electric energy (Reference 1)	Installed capacity, 1970
	MW
World	19060
Canada	11*
France	240
GDR	75†
W. Germany	958
India	420
Italy	642
Japan	1336
Netherlands	54
Spain	153
Sweden	10
Switzerland	350*
USSR	1150†
UK	4813
USA	6493

* 1969 figures

† 1968 figures

After about 100 years, the main radioactive materials of importance are strontium-90 and caesium-137, which will be required to be stored for between 600 and 1000 years. In addition, the waste contains small quantities of transuranic elements, particularly plutonium-239. This element has a halflife of $2 \cdot 5 \times 10^4$ years, and, because of its toxicity, it will be necessary to store for at least $2 \cdot 5 \times 10^5$ years.

Various approaches to the problem of fission waste have been adopted. In the UK, the waste is stored in liquid form in storage tanks, but studies are in progress to fix the waste in glass and load into stainless steel drums, which are to be stored in mine shafts. Initially, some cooling will be required. In the USA, the practice has been to store in liquid form in surface tanks, though there is now considerable interest in the possibility of storage in salt deposits. Longer-term solutions that have been suggested include sinking in the antarctic ice (Reference 12), and even firing in rockets at velocities greater than the escape velocity from the solar system. The cost of fission-waste disposal is a small fraction of the total cost of nuclear power, less than 1%, so that there is a considerable margin for improved disposal before the cost becomes significant. One possibility that should be considered is the separation and re-irradiation of fission products to reduce the transuranic content. Low-level waste is disposed of by controlled release to the environment. As the total release of radioactivity increases, the possibility of concentration by biological chains may become important and require additional measures to be taken.

2.3 Nuclear-fission energy resources

Table 4 lists the total uranium reserves that can be extracted at less than 10 $/lb as 1 575 000 t. This material, if burnt in a thermal reactor, would produce about the same energy as the oil reserves. As Hubbert has pointed out, there is no shortage of uranium in less rich concentrations. Hubbert[3] shows that, in the USA, the Chattanooga shale, which occurs in a layer 15 ft thick and extends over an area of hundreds of square miles, contains about 60 g/t of uranium. Since 1 g of uranium is equivalent to 2·7 t of coal, if the density of the shale is 2·5 t/m^3, one square metre of surface is equivalent to 2000 t of coal or 10 000 barrels of crude oil. Hence an area only about three miles square is equivalent in energy terms to the whole of the US crude-oil resources. Similar calculations have been made for Conway Granite in New Hampshire. However, although there is undoubtedly a considerable amount of uranium, it is necessary to ask whether it can be extracted and burnt economically. The current thermal reactors convert so inefficiently (CR of 0·6 for light-water reactors and 0·8 for gas-cooled reactors) that they can be regarded as burners. Fuel cost is an important component of the cost of electricity generated, and they are therefore limited to cheap fuel of 10 $/lb. How serious is the drain on the cheap uranium resources is brought out clearly in Table 7 due to Benedict (Reference 13). Benedict also shows (Reference 13) the total uranium resources of the USA as a function of cost (Table 8).

Table 7

US uranium requirements and resources (Reference 13)

	Actual (1970)	Projected 1980	2000
Electric generating capacity, $\times 10^6$ kW			
Total	300	523	1550
Nuclear	6	145	735
Total tons uranium concentrates consumed	—	200 000	1 600 000

Now the situation is changed markedly if breeder reactors are introduced. The fuel utilisation changes from about 2% for light-water reactors to 70% for breeder reactors even when losses due to repeated recycling are taken into account. Hence much better use of fuel is made; more important, since the cost of electricity is almost independent of fuel cost in a breeder reactor, the more expensive ores can be employed (Table 9 Reference 13).

Table 8

US uranium requirements and resources (Reference 13)

Price of uranium concentrates	Tons of uranium resources at this or lower price	Increase in cost of electricity from water-cooled nuclear power plants.
$/lb U_2O_2	(Reference 5)	mills/kW h
8	594 000	0·0
10	940 000	0·1
15	1 450 000	0·4
30	2 240 000	1·3
50	10 000 000	2·5
100	25 000 000	5·5

The total fuel requirement and hence the cost of fuel, will depend on a number of factors including the type of breeder reactor adopted and the rate at which it is introduced. This is illustrated in Fig. 6, taken from Reference 4.

Nuclear fusion

Hydrogen exists in several isotopic forms, hydrogen $_1H^1$, deuterium $_1H^2$, and tritium $_1H^3$. The nuclei of hydrogen may be fused together to form isotopes of helium $_2He^3$ and $_2He^4$ with the release of energy. Such reactions are known as fusion reactions.

Deuteron-deuteron reactions occur, either as

$$_1H^2 + _1H^2 \rightarrow _2He^3 + n + 3\cdot2 \text{ MeV}$$

or, with about equal probability,

$$_1H^2 + _1H^2 \rightarrow _1H^3 + _1H^1 + 4\cdot0 \text{ MeV}$$

The tritium nucleus in the second reaction will react with a further deuteron as

$$_1H^2 + _1H^3 \rightarrow _2He^4 + n + 17\cdot6 \text{ MeV}$$

The net reaction is

$$5_1H^2 \rightarrow _2He^4 + _2He^3 + _1H^1 + 2n + 24\cdot8 \text{ MeV}$$

or a release of approximately 5 MeV per deuteron consumed. Neutrons are also produced in the reaction.

Tritium may also be produced by the bombardment of lithium $_3Li^6$ according to the reaction

$$_3Li^6 + n \rightarrow _2He^4 + _1H^3 + 4\cdot8 \text{ MeV}$$

The tritium will then react with the deuterium as before.

The net reaction is

$$_3Li^6 + n + {}_1H^2 \rightarrow 2\,_2He^4 + n + 22{\cdot}4 \text{ MeV}$$

To achieve a self-sustaining thermonuclear reaction, the temperature of the gas must be raised to around 10^8K. In addition, the product of the pressure and the containment time must exceed a minimum value. It is this condition that has not yet been achieved in the laboratory. The deuterium-tritium reaction should be more easily achieved than the deuterium-deuterium reaction. The total energy that can be released will be limited by the amount of lithium available, and will be similar to the energy content of the fossil fuels.

Table 9

Comparison of uranium requirements of water-cooled reactors and fast-breeder reactors (Reference 13)

	Water-cooled	Fast breeder
Principal raw material	^{225}U	^{225}U
Uranium-concentrate consumption, tons per million kilowatt years	171	1·3
Increase in cost of electricity caused by increase in price of uranium, mills/kWh/$/lb.	0·06	0·0

Uranium price	US uranium resources	Increase in cost of electricity		Million kilowatt-years of electricity that could be generated	
		Water reactor	Fast breeder	Water reactors	Fast breeders
$/lb	tons	mills/kWh			
8	594 000	0·0	0·0	3 470	460 000
10	940 000	0·1	0·0	5 500	720 000
15	1 450 000	0·4	0·0	8 480	1 120 000
30	2 240 000	1·3	0·0	13 100	1 720 000
50	10 000 000	2·5	0·0	58 300	7 700 000
100	25 000 000	5·5	0·0	146 000	19 200 000

The technical aspects of fusion reactors are discussed elsewhere.

3   Conclusions

The future role of nuclear power (Reference 14)

The technical feasibility of nuclear fission has now been conclusively demonstrated. Controlled thermonuclear fusion is still some way off, though there seems to be no reason for doubting that it will ultimately be achieved. There are sufficient energy resources for both fission and fusion to satisfy world energy needs in the foreseeable future, though, in the case of fission, it is urgently necessary to mount a large breeder programme in order that sufficient fissile material should be bred.

A major problem with the increased use of fission power is the disposal of fission waste. Radioactive waste is much less serious for fusion power, though the problem of tritium release must be carefully considered, together with that of disposal of induced radioactive substances resulting from neutron bombardment of structural materials.

Both nuclear fission and fusion power plants will be large central facilities producing heat. This heat may be used to produce electricity or used as industrial-process heat. A major energy need is for a transportable fuel; nuclear power is not directly applicable, except in the relatively restricted application of nuclear-propelled ships. Electrical storage using nuclear-generated electricity is a possibility, but the present state of electrical storage is not promising. An alternative is the use of nuclear heat in the gasification of coal to give synthetic fuels and later the direct synthesis from limestone and water to give hydrogen, methane and methanol.

4 References

1   United Nations Statistical Yearbook 1971

2   COOK, E.   Sci. Amer. Sept. 1971, p.135

3   KING HUBBERT, M.   'Resources and man' (W.H. Freeman, 1969),
                                      chap.8

4   CULLER, F.L., and   'Energy from breeder reactors', Phys. Today,
    HARMS, W.O.   May 1972

5   'Nuclear power and the environment' (International
                                      Atomic Energy Agency, 1972)

6   FARMER, F.R. and   'Reactor safety philosophy and experimental
    GILBY, E.V.   verification'. 4th United Nations Conference on
                                      the peaceful uses of atomic energy, Geneva,
                                      Sept. 1971

7   SLATER, H.G.   'Public opposition to nuclear power: an industry
                                      overview'. Nucl. Safety, Sept.:Oct. 1971. **12**, (5)

8   DALE G.C., and   'Safety in nuclear power plants'. 4th United
    HARRISON J.R.   Nations conference on the peaceful uses of
                                      atomic energy, Geneva, Sept. 1971

9   FORD, D.F. and   'A critique of the AEC's interim criteria for
    KENDALL, H.W. and   emergency core-cooling systems', Nucl. News,
    MACKENZIE, J.J.   Jan. 1972, p.28

10 RASMUSSEN, N.C.   'Nuclear reactor safety — an opinion', ibid.,
                                      Jan. 1972

11 FARMER, F.R.   'Safety assessment of fast sodium-cooled
                                      reactors in the United Kingdom', Nucl. Safety,
                                      11, (4), July-Aug. 1970

12	ZELLER, E.J. SAUNDERS, D.F. and ANGINO, E.E.	'Putting radioactive wastes on ice', Bull. Atomic Scientists, Jan. 1973, **24**, (1), p.4
13	BENEDICT, M.	'Electric power from nuclear fission', Proc. Nat. Acad. Sci., 1971, **68**, (8), p.1923
14	WEINBERG A.M. and HAMMOND R.P.	'Global effects of increased use of energy'. 4th United Nations conference on the peaceful uses of atomic energy, Geneva, Sept. 1971

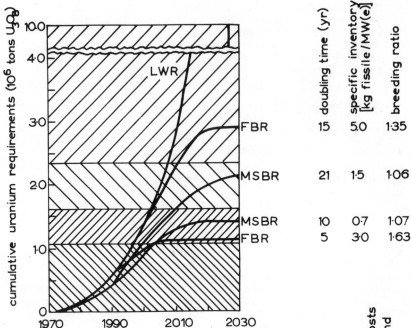

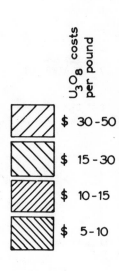

Fig. 6 U_3O_8 costs

2.2.4(i) Energy and the environment

Prof. H. Alfvén
Professor of Physics, University of California, USA.

Is there an energy crisis? It seems that the 'crisis' is much èxaggerated. At least·in part, the very rapid increase in energy consumption is due to 'promotional rates'. A distinction must be made between real needs and market considerations. But even if one cannot speak of a crisis, our technological culture depends to a large extent on an abundance of energy. It is highly desirable to get more energy, specially for the less developed countries. Also, in the industrialised countries, a certain increase in power supply is desirable to the extent that this can be achieved without too much insult to the environment. Thus the main real argument against continuing to burn fossil fuel is the increased concern about pollution.

The price of different kinds of energy is important to the consumer, but it is not equally important for the planning of energy policy. The reason is that the price reflects the real costs only to a limited extent. It is mainly decided by the policy of governments and private companies. Hence introducing the present (or future) price of different fuels into a discussion of energy policy easily leads to circularity in reasoning. This is specially applicable to the price of fission energy, which is made attractive by a low price. This price, however, does not include the development costs which have been charged to military funds in connection with the production of bombs, or which will be left to future generations to take care of in dealing with radioactive wastes.

If methods for purifying fuel and cleaning the exhaust of gases are successfully developed, we can meet energy needs adequately with fossil fuels for a period long enough to plan and deploy a sensible world policy for the energy supply and for developing new energy sources. The main new energy sources that sooner or later must take over are: fission energy, fusion energy and solar energy. Whether geothermal energy is abundant enough to be of major importance is not quite clear. Waterfalls and tidal energy will not be very important.

Fission energy

The energy generation process in a fission reactor leads to production of plutonium. (In some reactors, plutonium is burned.) In all reactors, the quantities of plutonium are of the order of kilograms or more. The energy generation processes necessarily lead also to the production of a number of other radioactive substances, among them strontium 90.

Both plutonium and strontium 90 are among the most poisonous substances we know. Under certain conditions, a few micrograms (one-millionth of a gram or one-billionth of a kilogram) is lethal: It produces cancer. The quanity of these substances in one ordinary reactor is of the order of a billion times the lethal dose. Hence the acceptability of fission energy depends on whether it is possible to keep these substances under control so that human beings are efficiently screened from them.

Unknown processes

In the biosphere, there exist a number of complicated processes that may enrich both plutonium and strontium (and other radioactive substances) — for example, in some common foodstuffs. Several of these processes are known, but there are reasons to suppose that there also are a great number of still unknown processes. Hence plutonium, strontium (and a number of other radioactive substances) must be efficently separated from the whole biosphere. Only very small quantities can be allowed to leak out. They decay automatically but some of them so slowly that the radioactive waste must be kept isolated from the biosphere for hundreds or thousands of years.

During the whole development of fission-reactor technology, the constructors have been aware of the importance of an efficient confinement of radioactive substances, although not until recently has the extremely dangerous nature of these substances been fully clarified. An enormous amount of highly qualified work has been devoted to insulating the poisonous substances.

It is claimed by the reactor constructors that, under normal operating conditions, a fission reactor is no insult to the environment. This is questioned by some environmentalists. Because of the enormous complexity of the problems, neither side can prove its views definitely. From the discussion, one gets the impression that the claim of the reactor constructor — that the reactor under normal operating conditions is not harmful — is likely to be correct. If we accept this view, it is appropriate to express great admiration for this achievement. Nobody who has visited a reactor station can avoid being deeply impressed by the ingenuity and skill that are manifest in the safety precautions.

The waste problem

But the fission reactor represents only one part in a complicated series of operations for fission-energy production. When the fuel elements are burnt out, they are taken out of the reactor, transported to a fuel processing plant, and later transported back again to the reactor. The processing plant extracts the highly radioactive waste, which is stored and finally placed in 'repositories', where it should remain indefinitely. The waste consists of extremely poisonous radioactive substances, which must be kept isolated from the biosphere until they decay after centuries or millennia.

The fuel-processing plants have not received so much attention in the discussion. It seems not to be clear to what extent they are dangerous; but, at least for the moment, it can be assumed that, under normal conditions, a processing plant is reasonably safe, or at least can be made safe without too much effort.

The handling of the waste seems to be a highly controversial problem. It is theoretically possible to transmute the radioactive substances into nonradioactive — and hence innocent — elements by nuclear reactions. If this were done, the waste problem would be solved. But such operations are complicated, and probably very expensive. If this solution were chosen, fission energy would lose the economic attractiveness that it now has, at least formally. (It has been suggested that the waste products should be shot out in space. Space would be a safe repository, but to shoot them out is extremely expensive. Moreover, if a rocket fails, there is a serious risk of atmosphere poisoning.)

According to present plans, the waste is to be deposited in salt mines. It is claimed by some that these are geologically so stable that there is no risk of leakage from the repository into the biosphere. This is questioned by a number of geologists. There is no doubt that the salt mines could be considered safe for any normal waste products. But because of the very large quantities of extremely poisonous substances, it is required that the repository should be absolutely free of leakage for periods of hundreds or thousands of years. No responsible geologist can guarantee this, simply because the problem is one of which we have no experience.

Key issues

We now approach the key issues. The reactor constructors claim that they have devoted more effort to safety problems than any other technologists have. This is true. From the beginning, they have paid much attention to safety, and they have been remarkably clever in devising safety precautions. This is perhaps pathetic, but it is not relevant. If a problem is too difficult to solve, one cannot claim that it is solved by pointing to all the efforts made to solve it.

The technologists claim that if everything works according to their blueprints, fission energy will be a safe and very attractive solution to the energy needs of the world. This may be correct. Hence they consider all the objections to be due to 'ignorance', 'viciousness' or 'hysteria'. This is not correct. The real issue is whether their blueprints will work in the real world and not only in a 'technological paradise'.

Fission-energy advocates say that, in all other technologies, one accepts certain risks, and that there is no way of completely eliminating all hazards. They claim to have taken all reasonable precautions to eliminate risks. What more can they do? This may very well be true, but it is irrelevant, because we are facing risks the nature of which we have never before experienced. The consequences of nuclear catastrophes are so terrible that risks that usually are considered to be normal are unacceptable in this field.

Opponents of fission energy point out a number of differences between the real world and the 'technological paradise'. Fission energy is safe only if a number of critical devices work as they should, if a number of people in key positions follow all their instructions, if there is no sabotage, no hijacking of the transports, if no reactor-fuel processing plant or repository anywhere in the world is situated in a region of riots or guerrilla activity, and no revolution or war — even a 'conventional' one — takes place in these regions. The enormous quantities of extremely dangerous material must not get into the hands of ignorant people or desperados. No acts of God can be permitted.

Advocates of fission energy have not given any reasonable answer to these objections. It is difficult to see how a satisfactory answer could be given. But if it cannot, we have to conclude that fission energy does not represent an acceptable solution to the energy problem. It will place an unendurable burden on the safety and health of future generations.

What has been said here applies both to the present-day fission reactors and to the projected breeders. In general, the breeders are much more dangerous, and current plans to develop breeders should be revised.

Fusion energy

A problem that was not discussed in the 'Energy crisis' articles is the relation between the peaceful atom and its militant twin brother. This is an extremely serious problem, even more important than the problems discussed here; it would, however, carry us too far to discuss it here.

Although a fusion reactor cannot be constructed today, enough is known of the basic processes of fusion and plasma physics to discuss the general properties of future fusion reactors.

In contrast to the fission process, the fusion processes that release energy do not result in radioactive waste products. In principle, therefore, fusion energy is 'clean', in the sense that it is not necessarily associated with the production of radioactive substances. The nuclear processes in a fusion reactor produce radiations, but these can easily be screened so that they do not reach the environment.

In case of a reactor accident, fusion reactions stop immediately by themselves. In the reactor, there are radioactive intermediate products which theoretically may escape, but, as the total quantities of radioactive material are very small in a fusion reactor, compared with a fission reactor, the danger is much less. Moreover, the radioactive products (mainly tritium) are much less dangerous to the biosphere.

Two problems

Owing to the intense neutron radiation inside a fusion reactor, the structure of the reactor may become radioactive. But, because there is a certain choice of structural material, the most dangerous elements can be avoided.

To construct a fusion reactor, two difficult problems must be solved: the confinement of a plasma and the transfer of the energy of swift neutrons. The first problem requires a general development of plasma physics. Very substantial progress has been made in this area, and we expect that the problem will be solved in the near future. The second problem is related to the handling of neutrons in the breeder; this also is likely to be solved in the near future.

What the 'near future' means depends on how much effort is devoted to research and development in this field. If the construction of a fusion reactor is considered to be very urgent, ten years may suffice. There are four different lines of approach that attract most interest: Tokamak, ring configurations, mirror machines with direct conversion, and evaporation of pellets. It is very likely that at least one of them will be successful.

As the fuel is heavy hydrogen, which can be obtained from ordinary water, there is enough raw material available in all countries. Thus, the fusion reactor would be an 'almost clean' long-term solution of the energy problem.

Solar energy

Solar-energy convertors produce no wastes. Large areas of land with sunny climate have to be used to collect solar radiation. If desert areas are used, the insult to the environment may not be considered serious.

Solar energy is available now, but at a price that is some hundred times the ordinary price of energy. There are a number of different ways of collecting solar energy: by mirrors, plastic lenses, 'hot-house' arrangements, photoelectric cells, and so forth. A research and development effort will certainly reduce the price, but the attainable level is impossible to predict at present.

Solar energy is available preferentially in countries with a sunny climate. The power is about $1\,GW/km^2$ during sunny hours. Storage of energy for periods with no sunshine complicates its use. There is also the possibility of using solar energy normally collected by and stored in the seas ('sea-thermal' energy).

'Geothermal' energy is already in use as a locally important energy source in some countries, for example, Iceland and the Soviet Union.

Looking at the geographical distribution of energy sources, we can summarise as follows:

(a) Fossil fuels, uranium deposits and sunshine have a very uneven geographical distribution.

(b) Fusion energy will be available to all countries, but, as it will require much scientific and technological knowledge, the industrialised countries will have an advantage.

(c) Access to energy sources is so important to our technology that the needs of all countries must be satisfied to a reasonable extent.

(d) The struggle for energy sources is a major factor in world policy. To make possible a balanced and peaceful development in the world, energy problems must be handled by an international body.

Our most immediate need is to solve the problems of using fossil fuels with a minimum of environmental insult. Methods of purifying fuels and clearing exhaust gases should receive high priority.

Clarification is necessary with respect to the extent — if any — to which fission energy is a realistic solution of the energy problem. New suggestions of how to avoid the fatal consequences of the use of fission energy should be considered. A strong effort should be made to develop fusion and solar energy. Environmental consequences should be carefully studied.

Methods of transporting energy (through tankers, pipelines, electric transmission lines, or other means) should also be developed.

One world problem

The energy problem should be approached as one world problem: how to satisfy reasonable needs of all nations. It should not be treated as a series of national problems: how to satisfy the energy demand of our own nation even at the expense of other nations.

The possibility of depoliticising the world energy problem deserves special attention. An international institute ('world energy agency'), as far as possible independent of pressure from government and economic interests, is necessary. (In this connection, the Pugwash movement, which already has devoted much attention to these problems, has considerable responsibility.)

To build up such an institute would be a very important task. It is possible that the International Atomic Energy Agency (IAEA) could serve as a nucleus: it has done competent work in the field of fission energy and devoted considerable interest to the problems of fusion energy. It also has had unique experience in handling relations with the governments of different countries. An international institute of the SIPRI (Stockholm International Peace Research Institute) type may also be considered. The proposed institute should study the needs of energy in different countries, compare possible ways to satisfy this need, and plan a world energy policy, in a rational and at the same time realistic way.

2.2.4(ii) Scientific assessment of radiation hazards

Prof. J. Rotblat

Professor of Physics, Medical College of St. Bartholomew's Hospital, London, UK.

It is almost exactly 30 years since the first nuclear reactor went critical and nuclear energy became a reality. The scientists who set up the first nuclear reactor were actually involved in a project with a different aim, to make the atom bomb; but, no doubt, many of them though that, as a compensation for the introduction of nuclear weapons, they were giving humanity a new source of energy for peaceful purposes, a source that would be inexhaustible, cheap and clean.

After the experience of 30 years, we have to conclude sadly that none of these promises has come completely true. Nuclear energy based on fission, although abundant, is certainly not inexhaustible; indeed, with the projected demands, it may be exhausted quite soon unless we are prepared to pay the heavy costs for mining uranium ores of low concentration; or unless we change to breeder reactors, which as yet have not been tested on a practical scale. Fission energy is also neither cheap nor clean. These two factors are somewhat connected, because a considerable proportion of the cost of nuclear reactors is on the prevention of accidents and the release of radioactivity, and this requires the taking of precautions far greater than in conventional power plants. However, even with the best design, some radioactivity will be released, and this raises the question of the pollution of the environment, about which many people are now deeply concerned.

The reason we are more concerned about these problems now than 30 years ago is that, as time went on, we have learnt more and more about the harmful effects of radiation, and the need to avoid exposure, even to very small doses.

I said that we have learnt a great deal about the effects of radiation, but, even with all this new knowledge, there are still huge gaps. One result of our ignorance of the relevant effects of radiation is that the assessment of the possible harm that could result from exposure to radiation ranges over orders of magnitude. For example, if one tries to calculate how many cases of cancer might result from the widespread use of nuclear energy, we find in the literature, figures varying from two per century to over 10 000 per year in the United Kingdom. Not all of this disagreement is due to our lack of knowledge about the effects of radiation; it is due to two reasons: one is the different assumptions made about the possible effects of small doses of radiation, and the other is the different assumptions made about the actual doses, the quantity of radiation to which the population may be exposed; but together these two factors give this enormous range of uncertainty.

What have we learnt in the recent years about the effects of radiation? In the short time available to me, I can only take up certain aspects, and I am going to deal mainly with the so-called somatic effects, the effects on the individual exposed; of these, our main concern is about the induction of cancers.

The first type of cancer that we knew to be induced by radiation was leukaemia. For a long time, it was thought that only exposure to high doses of radiation causes leukaemia, and even now some people claim that there is a threshold, a safe dose, below which no harm will come at all. However, recent surveys made on the survivors of the atom bombs in Hiroshima and Nagasaki show conclusively that the limit of 100 rad, which some scientists believed to be the dose below which one is safe, does not exist. In this survey, the incidence of leukaemia among the survivors is plotted as a function of the dose of radiation estimated to have been received by them. For Hiroshima, the graph shows a definite increase for the group that received a dose between 20 and 50 rad. Of course, at higher doses, the incidence of leukaemia goes up rapidly. There seems to be a difference between the effects observed in Hiroshima and Nagasaki; this may be possibly due to the different types of bomb – in one case, there were more neutrons, and, in the other, almost entirely gamma rays, and these two types of radiation may give different effects. However, if one carries out a mathematical analysis of the data, one finds that both fit in quite well with linear relations, which indicates that any increase in the dose of radiation is likely to produce an increase in leukaemia. I do not think that one can doubt any more that this is the case.

The next point is that, at one time, many scientists believed that only leukaemia can be induced by radiation when the whole body is exposed, and that other types of cancer are not induced by radiation unless very high local doses are given. This too we now know not to be true. The reason for this misconception is that leukaemia is the first cancer to appear after irradiation; it comes within a few years, whereas other cancers take much longer – 10 to 30 years – to develop. This came to light as the result of a survey on a population that was exposed to radiation for the treatment of a disease of the spine – ankylosing spondilitis. The followup of these patients over many years after the exposure shows clearly that the incidence of leukaemia started very early, grew over several years, and then began to decline, while, by contrast, other cancers, which showed a low incidence in the early years, are now gradually increasing, and, even after 27 years, the incidence of all sorts of cancer seems still to increase compared with unirradiated populations. It seems now practically certain that almost every type of cancer that is known to occur without radiation can be induced and its frequency increased by exposure to radiation.

The third point was about the special sensitivity of the foetus to radiation. It was found that, if mothers received ordinary X rays for diagnostic purposes during pregnancy, the children born to them would have a higher probability of incurring leukaemia and other cancers than children whose mothers were not X rayed. Moreover, by studying the excess of cancers in the children born to X rayed mothers as a function of the number of films taken, and, assuming that each film gave about the same dose, one can deduce that here too there is a linear relation between incidence of cancer and radiation dose. However, in this case, these doses are very small, of the order of 1 rad. Thus you see that, in the case of the foetus, even very small doses can induce leukaemia and other cancers.

These are the main new facts that have emerged in the course of recent years, and which have added to our worry about the possible effects of radiation. This greater awareness is reflected in the recommendations that the official bodies regulating the exposure to radiations have been making over the years, particularly for people who are occupationally exposed to radiation – workers in the various radiation industries – how much they can safely, or with a very small amount of risk, receive. This is called the maximum permissible dose (MPD). Table 1 shows the value of the MPD as a function of time since 1924. You see that, at the beginning, we were allowed 1·5 rad per week. Then the allowed dose has gone down and down, and the value in force now is only 0·1 rad. This is what we are allowed now, and there has been no change since 1956. The reason the MPDs kept going down is that, as time went on, we found out more and more about the long-term effects, i.e. induction of cancers, which, as you have seen, takes years to appear. My own feeling is that the time has come to take another plunge and reduce the maximum permissible dose by a factor of 10. We often hear that radiation work is one of the safest industries, that because we can measure and monitor the radiation dose received by workers accurately, we can estimate the hazard and ensure that it is very small compared with other industries. I submit that, in the light of the most recent findings, this is no longer true. Table 2 lists the probability of a worker dying as a result of his work, for various industries. It is seen that a radiation worker, if exposed to the MPD level, has a death risk of 2%, whereas the average risk for most other occupations in Great Britain is only 0·3%. Therefore radiation no longer ranks among the safe occupations; it has become one of the more hazardous occupations. This is one of the reasons I feel that the MPDs should go down by an order of magnitude. I should add that very few people really get the MPD. The actual dose received by workers is less than 10% of the MPD. However, as long as the present MPDs stand, nobody exposed to a dose up to the MPD can legally claim any compensation, should he suffer any harm.

Table 1

Maximum permissible doses (MPD) for occupational exposure

Year	rad per week
1924	1·5
1934	1·0
1950	0·3
1956	0·1

Table 2

Risk of death in selected occupations in the UK

Occupation	percentage
Trawler fishing	5
Aircraft crews (civilian)	7
Coal mining	2
Pottery (pneumoconiosis)	2
Construction	5
All manufacturing	0·3
Radiation workers (if exposed to MPD)	2

The same applies to the general population. The dose to the population is supposed to be limited by genetic considerations, and the so-called genetic dose is 5 rad per 30 years. Assuming a linear relationship, and with the figures that we know now about cancers, I have calculated the possible number of cancers that may occur in the whole population if exposed to the genetic dose, 5 rad in 30 years. On the basis of these figures, one might expect something like 600 additional cases of cancer per year in this country, and 2500 per year in the USA. Once again it should be stated that the population in general is not exposed to the whole of the genetic dose, we do not get anything near 170 mrad per year; indeed the population dose due to nuclear power is at present less than 1 mrad per year, and this is why the protagonists of nuclear energy keep assuring us that we do not need to worry. However, other people – the antagonists of the nuclear-energy programme – are much less sanguine; they say that, as long as regulations are in existence that permit a higher dose, and if the need to keep to a smaller dose may make nuclear energy much more expensive, because of the high cost of the more stringent safety precautions, it is possible that the industry will relax its standards and allow the dose to go up to the limit.

This is certainly the view taken by two of the strongest antagonists of the nuclear-energy programme in the USA, Gofman and Tamplin.

These two scientists worked in the Livermore Radiation Laboratory in California, which was run initially by Edward Teller, who does not believe in the harmful effects of small doses of radiation. When the treaty to ban nuclear tests came about, largely because public opinion felt very strongly about the harmful effects of radiation, Teller decided to set up a special biomedical division at Livermore to carry out research into the effects of radiation, perhaps in the hope that this research would prove that the effects have been grossly exaggerated. Prof. Gofman was the head of this division, and Dr. Tamplin was working with him. Recently they came out with public statements that suggest that the effects are far worse than anyone of us had ever imagined. Gofman's and Tamplin's philosophy is the so-called 'worst case' approach; they claim that, when it comes to human life, or the health of human beings, one must always take the most cautious approach. If one is uncertain or in doubt, one must assume that the worst might happen. They applied this philosophy to the nuclear-energy programme, and, on this basis, produced figures about radiation risks that are well beyond the ones accepted by the general community of scientists, and on which I based the figures that I gave before.

Gofman and Tamplin have analysed the incidence of cancer due to radiation exposure, and, taking in each case the worst figures, they come to the conclusion that the doubling dose – the dose required to double the natural incidence of cancer – is about the same for all cancers, and is only about 50 rad. If 50 rad doubles the number of cancer deaths, a dose of 5 rad will produce an increase of 10%, which for the USA would mean an additional 32 000 deaths from cancer every year. Recently, they have refined their calculations, and now claim that, in fact, there may be 104 000 additional deaths from cancer every year in the USA if the population were exposed to the genetic dose.

I should mention that these calculations are based on a new and very important concept that is not accepted generally. The concept of a constant doubling dose means that the same dose, say 1 rad, may produce a small number of cancers, which are naturally rare, such as leukaemia, but a much larger number of cases of cancers that are more prevalent, such as lung cancer. This suggestion, that the same dose may give different numbers of cancers depending on whether the type of cancer is rare of frequent, is not easily acceptable. Why should it be so? Why should people who smoke, and therefore have a much greater chance of getting cancer of the lung, be more susceptible to radiation than people who do not smoke? Gofman and Tamplin explain this by synergism, that is to say that radiation acts synergistically with other causes of cancer. In his paper, Prof. Rosenquist dismisses the additiveness of radiation and other sources of environmental pollution, but Gofman and Tamplin go much further; they say that the result is more than additive, it is synergistic. In other words, if a person is exposed to two causes simultaneously, the result is much greater than the sum of the two individual factors. They give figures to prove their case, based on the observation of cancer of the lung in uranium miners, some of whom were smokers and others who were not smokers. This evidence is open to criticism, but, if they should prove to be right, this would be of great importance in a world in which we are exposed to so many pollutants. It is believed that about 90% of all cancers are manmade, maybe even 100%. If, therefore, there exists a synergistic action between different causes of cancer, any additional source could be much more harmful than might have been thought before. This is why I think that the views of Gofman and Tamplin warrant careful attention, and that much more research needs to be done on these problems.

Returning to radiation levels, Table 3 shows the dose the average person receives in this country, to the bone marrow, that is relevant to the induction of leukaemia. It is seen that the natural background is about 90 mrad per year. As far as manmade sources are concerned, the biggest one comes from medical diagnosis. At the present time, this accounts for 32 mrad. Tests of nuclear weapons produce at the present time a dose of about 5 mrad per year. Various industries, including the atomic energy industry and the uses of radioactive isotopes, give a dose of the order of 1 mrad. Thus, at the present time, this represents a very small addition to the total, and therefore, on this basis, one need not worry too much about the nuclear-energy industry.

Table 3

Bone-marrow dose to the population from various sources

	millirad per year
Natural background	90
Medical (diagnostic)	32
Fallout from tests	5
Various industries	1

However, that is the position at present. But we are talking about a possible increase of nuclear-energy production by a factor of 200 by the year 2000, and this will create an entirely different situation.

There are at least four important ways in which large-scale nuclear-energy production may release radioactivity. One is from the planned release of radiation from reactors. In every reactor, a certain amount of radioactivity escapes, because the fuel elements cannot be sealed off completely and a certain amount of the fission fragments come out. However, this escape can be made quite small. Recently, the US Government decided to reduce the allowable release of radioactivity from reactors by a factor of 100. More difficult will be to tackle the planned release from reprocessing plants. The fuel elements have to be taken out from the reactor from time to time and processed to obtain the plutonium accumulated there, as well as to remove the radioactive fission fragments. In this process, a certain amount of release occurs, mainly of gaseous products, krypton and tritium; these are the most important elements that come out. Here again, by proper design of the plant – and if cost is not a major obstacle – one could reduce this release as well. Therefore the exposures from planned releases could be kept at low levels, not safe levels, but low compared with the other doses to which we are being exposed.

Of the other three possible releases, two have been mentioned by Prof. Alfven; one is a possible accident to a reactor, and another the storage of long-lived radioactive wastes. The one that he has not mentioned is a possible release of radioactivity in an accident in transportation. One has to transport enormous quantities of radioactive material from the reactors to the processing plants; they have to be somehow conveyed by rail or by road, and, in the course of this transportation, accidents may occur and release radioactivity into the open.

These three types of hazard, resulting from accidents to the reactors themselves, accidents in transportation, and difficulties of storage, are not the sort of events that lend themselves to easy evaluation. Since the title of my paper is the 'Scientific assessment of radiation hazards', I dare not make a guess about the magnitude of these hazards, simply because one cannot produce any reliable figures. All we know, as Prof. Alfven has said, is that nothing can be made absolutely foolproof, that sooner or later an accident will occur. What we can do is to take measures that, if an accident does occur, it will be not on a very large scale. On this basis, we should be wary about large breeder reactors, because they appear to be inherently less safe than thermal reactors, and an accident in them would be on a much larger scale than in ordinary reactors.

Storage of long-lived radioactive waste is a formidable problem. It is calculated that, by the end of this century, if we proceed with the nuclear-energy programme as envisaged now, there will be something of the order of 30 GCi — 30 thousand million curies of radioactivity produced from nuclear reactors. The heat released spontaneously as a result of their natural decay corresponds to a power level of 100 000 kW. This is just the heat released from the products themselves, and nothing can be done to stop it. If we do not dissipate the heat by cooling, the radioactive materials will melt any container and contaminate the environment. Therefore something must be done to remove them from our daily environment. It has been suggested that they be put into rockets and fired into space. However, with the present technology of satellites, we cannot guarantee that they will not come back to us! Another means of disposal is by putting them in deep disused mines, preferably salt mines, but here too one cannot be certain that they will not seep through and get into the water system. Up to now, there is no definite solution to the problem of wastes. Considering that their radioactivity will last for many centuries, we may be justified in questioning the morality of embarking on a venture that will leave a dangerous situation to future generations and commit them to tackle a problem that we cannot solve.

2.3.1 Brief survey of the history and present state of knowledge of the use of windpower

J. R. Tagg.
Faculty of Technology, Open University, UK.

For perhaps 4000 years, man has used windmills to supplement his own labour. During the last millenium, the most successful machines have had blades or sails arranged to rotate in a vertical plane about a horizontal axis. The number of blades in the wind rotor has varied from two in the higher-speed machines to, perhaps, 24 or more in the slow-speed mills intended for pumping water or grinding corn.

In most designs, sufficient ground clearance is obtained by mounting the windshaft, carrying the blades, on the top of a tower. This is usually free to turn, so that the blades may be brought to face winds blowing from different directions. Alternatively, the whole tower can be arranged to rotate on a low-level pivot or post.

The early machines had an adjustable covering of cloth on the blade structure. Frequent stops were therefore necessary to adjust the covered area exposed to the wind to match its strength. Without such adjustments, there was an ever present danger that rising winds would cause a mill to over speed and destroy itself.

Starting in 1745, a succession of improvements by eminent British engineers, notably Lee, Meikle, Smeaton, Hooper and Cubitt, completely transformed the corn-grinding mill. By 1810, it worked automatically, and was probably the world's first automatic factory. The blade setting was controlled to ensure that work started as soon as the wind was sufficiently strong. The sails then continued to turn at a steady speed in rising winds. In stronger winds, the rotational speed was held at a safe low level by the automatic opening of slats under the influence of wind pressure. A fantail wheel, with blades set at about 45° to the wind flow, was used to turn and hold the wind wheel facing the wind.

Within the mill, the corn being ground was automatically fed to the stones at a controlled rate from hoppers mounted aloft. The gap between the stones was adjustable, and was controlled by a governor to match their rotational speed. This regulated to some extent the power being absorbed from the wind. An alarm bell warned the miller when the amount of unground corn was running low, and he could then leave the machine to work unattended for long periods.

English cornmills competed successfully for some time against the newly introduced steam engines fuelled by cheap coal. But, eventually, the ability of external- and, later, internal-combustion engines to work continously whenever they were called on to do so gradually brought about the decline of the corn-grinding mill.

Water-pumping windmills continued in widespread use for some time, although their popularity too eventually declined. They had an advantage over the cornmills, in that, although the wind might not blow continuously, the pumping mill could be relied on to lift fairly uniform volumes of water over regular periods of time, always provided that there was sufficient water-storage capacity at either the input or the output to the pump.

This basic fact highlights the problem common to all wind-driven machines. Except in a few given situations, firm power from the wind cannot be guaranteed at any particular time or place. Nevertheless, over extended periods of, say, a year, the amount of energy that may be extracted from the wind at any given site will be substantially constant within a consecutive period.

During the early part of the 20th century, development of windmills for grinding and pumping continued sporadically overseas. Bilau in northern Germany and Dekker in Holland were both concerned with improving machine efficiency. An increase in annual running time with improved power outputs was achieved by covering the blades with sheet metal. Running friction was reduced by fitting roller bearings to the wind shaft, and steel became increasingly used in the construction of the blades. The towers, however, continued to be made largely of timber or brick.

In Denmark alone, the development of wind-driven machines had followed a different line. Here, around the end of the 19th century, attention had increasingly been focused on smaller machines developing up to 5 hp. These were built in large numbers from standard steel sections, and were intended for use on individual farms. Then, with financial support from the Danish government, La Cour initiated experimental work at Askov, which culminated in the construction of a much larger steel machine. Mounted on a lattice tower some 24m tall, the wind rotor had four shuttered blades, each about 12m long. The rotor speed was regulated by the shutters, which opened under the control of a governor if the rotational speed tended to increase. A conventional fan-tail wheel was used to hold the main wind rotor facing the wind.

Power extracted from the airflow was led through two sets of bevel gears and a vertical shaft to produce electricity from a dynamo mounted at ground level. By 1910, several hundred machines of this type were successfully supplying villages in Denmark with electric power. Maximum output varied between 5 and 25 kW at 110 or 220 V. Storage batteries of 100 to 300 Ah provided spare capacity to meet demand during calm periods. With fuel shortages accentuated by the Great War, additional machines were built. The largest of these generated up to 35 kW from a wind wheel of around 18m diameter.

After the war, electricity produced by diesel-driven generators became much cheaper, and, increasingly, attention was paid to reducing the capital cost of wind-driven plant. In France, Darrieus designed a light, 3-bladed machine, with the blades arranged to run downwind of the tower. This avoided the need for a fantail wheel to control orientation. A d.c. generator was mounted in a nacelle at hub height on the wind shaft. Its output was controlled by a simple regulator that adjusted the excitation. With this device, it was possible to restrict the range of rotor speed, thereby controlling the power generated. In effect, the regulation of speed and output power in rising winds resulted from an aerodynamic stall of the blades. This was the first time that such a method of power control had been used effectively.

Experiments continued in France for a number of years, and several machines were built. The largest produced 15 kW from a wind rotor of 20 m diameter in a wind of about 6 m/s. It was connected to the a.c. electrical network by a motor generator set, again the first time that such a connection had been attempted. In spite of the low energy costs achieved, the Darrieus machines were thought to be still too expensive for commercial exploitation.

Developments were taking place in Russia at the same time. Here, at Balaclava in 1931, a windmill driving an a.c. induction generator was arranged to feed energy directly into a local supply network. Acting as a negative load, the machine could be thought of as a fuel saver working only during windy periods. It was quite large, producing around 100 kW from blades of about 30 m diameter. These were mounted on a tower of a similar height. The tower was free to move round a circular track, and axial wind loads were resisted by a pivoted central stay. Output-power control was achieved by changing the pitch of the blades under the action of offset ailerons. It was reported that this was not entirely satisfactory, but nevertheless, some 280 000 kWh were generated in one year. This was at a site that had an average wind speed of around 7 m/s.

One of the most important developments took place in the USA during the Second World War. In 1941, the S. Morgan Smith Co built the largest ever wind-driven machine to the design of P.C. Putnam. The wind rotor had only two blades, each some 27 m long and weighing about 8 tons. They were mounted on a tower about 36 m tall. The unit was designed to produce a maximum output of 1250 kW from a synchronous alternator feeding directly into the local supply network. The whole project has been described extensively by Putnam in his book, 'Power from the wind' (Reference 1).

The unit suffered from an early bearing failure, and soon there were cracks in the blades owing to fatigue. After welding repairs had been carried out in the field, the machine ran successfully for some time. With only two blades, there were continuing problems of vibration, and the rotor was eventually destroyed by a catastrophic blade failure in 1945.

Grandpa's Knob, the site chosen for this experiment, was a 650 m hill in central Vermont. This was an area of, at that time, abundant and cheap electricity. In spite of the success achieved by the project up to the time of the blade breakage, it was decided to discontinue the experiment, and the machine was not repaired.

Meanwhile, there were more extensive, but less ambitious, developments going on in Denmark under the stimulus of another severe fuel shortage. Many old windmills were pressed back into service to produce electricity. The Lykegaard Co., following the principles laid down by La Cour, built many new machines. They were sufficiently confident of their reliability to install some of these without charge, merely relying on sales of electricity for a financial return.

Other, more advanced, designs were constructed by the Danish F.S. Smidth Co. The largest of these had three narrow aerofoil-section wooden blades, each some 12 m long. The wind rotor was coupled by a gearbox to a 70 kW d.c. generator mounted on the top of a reinforced-concrete tower, 24 m tall. Blade orientation was by a conventional fantail wheel. The energy costs from these machines were fully competitive with the costs of energy from local diesel-driven generators.

With the end of the Second World War, the development of wind power was restarted in Britain, France and Germany. In 1947, under the direction of Golding, staff of the Electrical Research Association began a series of surveys and experiments, which were to continue for over a quarter of a century (References 2-8).

At first, the ERA work was primarily concerned with a survey of the British Isles to demonstrate the potential wind energy available (Reference 9). With measurements proving that a suitable wind regime existed in the Orkney Islands, the first 100 kW windmill was erected there on Costa Head by the John Brown Co. This machine was fitted with an induction generator connected to the local electricity-supply network, and its intended role was, again, that of a fuel-saving device.

The Orkney windmill had three blades, each some 7·6 m long. They were fitted with universal joints at the blade roots, and this provided them with freedom to come forward under wind pressure, as well as giving them lateral freedom in the plane of rotation. The blades were also fitted with a hydraulic pitch-changing mechanism to provide a control of their speed. It was expected that, under the action of gusts, the blade system would generate both shock and oscillatory forces. The generator bed plate was therefore isolated from the 24 m tower by a system of sprung mounts. The windmill was run, and it produced electricity, but, for a variety of reasons, the design was unsatisfactory. Leakage of high-pressure oil feeding the pitch-changing mechanism posed a continual problem. The wind-rotor hub design was unnecessarily complicated.

The blades had too many degrees of freedom. Consequently, the machine was unstable, and the experiment was eventually abandoned.

A different approach was followed by Enfield Cables, who, in collaboration with de Havilland Propellors and the Redheugh Iron & Steel Co., built a 100 kW machine at a poor site near St. Albans. The design followed that of Andreau, and featured a pneumatic system of power transmission from the blades to a synchronous generator mounted near the base of the tower.

There were two hollow 12 m blades mounted at a hub height of 30 m. Their orientation was controlled by a vane-driven servomechanism. Air thrown out through the open-ended blade tips was replaced by air drawn up through the tower. In passing through a turbine , energy was extracted from it to drive a synchronous alternator connected to the local electricity supply. The system had an inherently low efficiency, but this was easily counterbalanced by a proportionate design increase in the length of the blades.

In passing, it should be noted that a high efficiency is not necessarily important in the design of a wind-driven machine. What matters is that the overall capital cost should be as low as possible for the rated duty. In the Andreau machine, the mechanical design was sound, but insufficient attention had been paid to minimising constructional costs. On analysis, these appeared to be too great by a factor of ten.

The machine had originally been intended for installation by the British Electricity Authority at a site in north Wales. But, with the expectation of cheaper electricity soon to be produced from nuclear fission, interest in the project slackened, and the windmill was shipped to Algeria. There it was installed on a windy site by Electricite et Gaz d'Algerie for parallel operation with their other generating plant. In 1961, it was reported that the Andreau windmill had operated successfully for extended periods, and that it had produced output powers of up to 180 kW.

The first two large British machines provided valuable operating experience. By the middle of the 1950s, it was realised that the successful development of wind power would depend on cheap but reliable machines installed on suitable windy sites. With the ERA acting as a catalyst, there were to be three more attempts in Britain to fulfil these conditions.

In response to support from the Ministry of Fuel and Power, the Dowsett organisation, designed and produced a machine. Leaning heavily on shipbuilding practice, a 10 kW machine was erected at Cranfield in Bedfordshire. It had a tapered tubular tower made out of rolled sheet steel. A fixed-pitch 2-bladed wind rotor drove an induction generator, and this was directly connected to the local electricity supply. Unfortunately, the lesson of the essential requirement of a fail-safe shut down and braking mechanism had not by then been learnt. While stationary and unattended, hydraulic oil pressure leaked away from the restraining brake. A light, gusty wind was enough to start the machine turning, and, in the absence of an electrical load, it accelerated rapidly to a very high speed. The rotor burst, and parts were distributed down wind for some hundreds of metres.

A second machine produced by the same organisation incorporated many improvements on the first. The tower, a sturdy tubular-steel-tripod structure, 10 m high, supported a 3-bladed wind rotor. This was arranged to drive a 25 kW induction generator through step-up gearing. As an economy, the generator was connected to the electrical network by a plug and socket with a trailing cable instead of the more usual slip rings. Orientation was by a conventional fantail wheel. (Fig. 1).

In the absence of sufficient wind, the blades were held in coarse pitch by heavy weights. Hydraulic oil pressure from a small pump was sufficient to raise the weights and turn the blades into a fixed fine-pitch running position, and the machine could then produce

Fig. 1   25 kW Dowsett under test at
the ERA windmill testing
station at Cranfield,
Beds, England

useful power in suitable winds. Mains failure, excessive wind speed or excessive rotational speed could all equally trip the oil pressure and return the blades to a safe coarse pitch.

Performance tests were completed successfully at Cranfield, which, by now, had become a world centre for the testing of wind-driven machines. The 25 kW windmill was shipped to the Isle of Man, where it was installed on a fairly windy site. There it continued to work well in parallel with the local diesel-powered electricity network. Other machines of the same type were built for isolated operation at remote sites, but the absence of profit from sales forced the manufacturers to withdraw from the market.

The Ministry of Fuel and Power were meanwhile sufficiently encouraged by the operation of the first 25 kW Dowsett machine to continue their support for the development of wind power. The Ministry commissioned Irwin & Partners to design, and R. Smith (Horley) Ltd. to build, a 100 kW machine on the same site in the Isle of Man, alongside the Dowsett machine. Mounted on a stronger 10 m tripod tower, the wind-rotor blades were made from three aluminium extrusions.

These were fitted into a forged steel hub, and were braced both forwards and sideways to withstand wind-induced loads. Again, a large fantail wheel controlled orientation, and trailing cables were used instead of slip rings for the electrical connection (Fig. 2).

Both machines in the Isle of Man eventually suffered from mechanical troubles. In the Dowsett machine, pins had been used to secure the fantail wheel to the pinion controlling orientation. These failed repeatedly, and allowed the head to rotate freely. Eventually, there was a serious gear failure in the gearbox, and the machine was scrapped.

The wind rotor of the larger machine was inadequately balanced. This, it was reported, set up a tendency for the head to rotate when the mechanical brake was applied. Since this brake acted on the high-speed gearbox shaft, airbrakes were fitted later to each blade tip as an additional safety precaution.

The airbrakes were initiated by separated shear pins, which were designed to yield when the rotor speed became too great. The pins gave trouble, because their operation could

Fig. 2   100 kW windmill built by
R. Smith (Horley) Ltd.

not be synchronised. In addition, the aerodynamic drag of the unopened airbrakes was unacceptably high. It was found that the alternator consumed about 30 kW when motoring under light-wind conditions. Eventually, a blade-stiffening brace broke, and this allowed the wind rotor to strike a tower leg.

Summarising the results of the Isle of Man experiments, it can be said that the initiators of the programme failed to foresee the commercial importance of using a windy site. They also failed to match the operating characteristics of the machine to the local wind regime. This work was finally discontinued, because it failed to yield an adequate commercial return on the investment.

During the same period, there had been encouraging parallel developments in Europe. Under the leadership of Dr. Juul, the South East Zealand Electricity Supply Co. of Denmark had constructed three machines (References 10-12). The first, a rather ugly experimental 13·5 kW windmill at Vester Eggesborg, was followed in the Winter of 1952-53 by an experimental 45 kW 3-bladed design. The wind rotor, 13 m in diameter, was mounted on a concrete tower at Bogø, 20 m high. Of moderate size, the design showed economic promise and the capability of being extrapolated to larger sizes. With both lateral and forward stiffening stays, the wind rotor was held upwind of the tower, and rotational speed was maintained sensibly constant by an induction generator tied to the network supply frequency. As a result of this electrical connection, the blades operated with a progressive aerodynamic stall. As with Darrieus, this limited the output power in rising winds. The starting-up procedure was automatic, and there were automatic mechanical- and air-braking arrangements to stop the machine in the event of a loss of electrical load.

Profiting from earlier Danish work, the third machine was erected at Gedser in 1957. With a wind rotor only 24 m in diameter; the same as the earlier 70 kW Smidth design, the new machine produced an output of around 200 kW. ERA staff co-operated closely in making stress analyses and performance tests of the new machine. It was found that the wind-induced stresses were commendably low, and the design seemed assured of a promising future (Fig. 3).

Fig. 3 200 kW windmill completed in Denmark in 1957

In Germany during the Hitler era, there had been a number of attempts to initiate large-scale projects. The names of Honnef, Kleinhenz and Teubert figure prominently in the literature (Reference 13). After the war, Hütter was responsible for the design of an outstanding small machine constructed in quantity by the manufacturing firm of Allgaier. Later, Hütter was to collaborate with Studiengesellschaft Windkraft e.V. and a number of German electricity-supply undertakings in developing a 100 kW design for direct connection to the electricity-supply network.

The smaller machine was engineered with typical German thoroughness and a fail-safe philosophy. Pitch changing of the 5m metal-covered blades regulated the output power according to the electrical loading on the machine. The blades were automatically turned into a safe low-speed coarse pitch by springs. This was initiated either by failure of the hydraulic oil pressure, by excessive blade speed or by the onset of high winds. Short of machine breakage, the design seemed faultless. The ERA operated an 8 kW version of this machine for many years in a priority-load-selection experiment at an isolated croft in Scotland.

At a recent workshop conference on windpower held by NASA in Cleveland, USA, the large 100 kW windmill was reported to have worked well. It featured an ingenious scheme for relieving blade stresses under wind loading. The blades were held rigidly in line with each other as a beam, but were allowed to pivot at the hub, either backwards or forwards into the wind flow.

In France, there had been similar parallel developments. Vadot, a consultant to the NEYRPIC organisation in Grenoble, had evolved a convenient 8 m diameter plastics-covered wind rotor for use in several configurations. He was to go on and design larger machines, including the one that was built at St Rémy des Landes (Manche) under the sponsorship of Electricité de France. It produced 132 kW from a wind rotor of 21 m diameter in winds of about 12 m/s. The blades, also plastics-covered, were free to change pitch and to furl completely in strong winds.

A more ambitious project was also commissioned by Electricite de France. Designed by their consulting engineer Lucien Romani, the machine was erected at Nogen le Roi (Eure et Loire). It had a wind rotor 30 m in diameter, made up of three fixed-pitch aluminium-alloy blades. Spoilers fitted to their leading edges operated under centrifugal force if the machine should overspeed. The blades were mounted at a hub height of about 32 m on a composite tubular-steel tower extending from the top of a 3-legged lattice structure. The designed maximum output power was 640 kW derived from a synchronous alternator driven at 1000 rev/min by a 2-stage step-up gear train. It was reported recently that the machine produced its full output in wind speeds of around 20 m/s.

At the time this report was being prepared, no final account of the performance of the two French machines had been received.

Reports of recent wind-power developments in Russia are rather scanty. Golding visited the country in 1961, and reported that some 50 000 water-pumping windmills had been constructed, and that about 30 000 were still in operation. They were generally either 3 m or 8 m in diameter. About 150 machines with rotors 12 m in diameter had been built to generate electricity at powers up to 10 kW. There had also been two larger machines fitted with 18 m rotors and producing 25 kW.

In a more recent report (Reference 14), Levy said that, in 1968, there were some 10 000 windmills in Russia. He listed the basic details of six types of machine that had all been designed for pumping water. In addition, Levy mentioned that, during the period 1935-55, some 30 prototype machines had been constructed for generating electricity at powers of about 30 kW. There had also been one 100 kW prototype. This was presumably the machine built at Balaclava and described earlier in the present account. Work was apparently still continuing. With the use of improved materials, the main emphasis was currently being placed on reducing capital costs, increasing the aerodynamic efficiency and extending the life of the machines. Other studies, he said, were devoted to improving the control systems.

In the early 1960s, the ERA was commissioned by the Egyptian government authorities to construct a small, 4 kW, windmill. Using a cable interconnection, the machine was designed to drive several small submersible bore-hole water pumps. These were to replace existing slow-running multibladed wind-driven pumps, which had been installed on unfavourable low-lying sites near the Mediterranean coast. By putting the new machine on an exposed hill top, it was expected that much more effective use would be made of the energy derived from the wind. Unfortunately, the design was based on inaccurate deductions of the likely wind regime. These deductions had been based on records from a meteorological station some distance from the new windmill site. It turned out to be far less windy than had been supposed, and the experiment was not successful.

At this time, financial support for wind power research could no longer be found in Great Britain, and ERA staff were forced to adopt a passive watching role. In 1971, their work received acknowledgment in the US Congressional Record for 7th December 1971, as follows:

'It appeared that Britain might step ahead with wind power just as she started to do with tidal power in the late 1940s but the nuclear scientists gained control. Britain opted for leadership in nuclear-fission central plants and won the race. Her work in wind power was continued by an able group within the Electrical Research Association, with Golding becoming a world leader in the technology. What had been a major thrust towards large-scale utilisation of wind power in the world's most heavily industrialised countries now became a matter of philosophic interest and a hope for energy-hungry emerging nations.'

The centres of interest in wind power now moved to Quebec and Barbados. Under the will of a Canadian engineer, the late Major James H. Brace, a special fund had been set up to ' . . . carry on research for the development of methods or means of eliminating or reducing the salt content of sea water so that it may be used economically and effectively for irrigation . . .' Under the terms of this bequest, the Brace Research Institute was established by the then retiring Dean of Engineering of McGill University Prof. R.E. Jamieson, and this led to the inauguration of a research centre in Barbados. The prototype Andreau windmill, which had been under test at Cranfield, was shipped out to the Barbados field station for experiments devoted to pumping water. After gaining experience with wind machines, staff at Barbados then turned their attention to improving the economy of wind-driven pumps by reducing their capital cost. Based on a wind rotor of three 5-metre glass-fibre blades, the new machine was arranged to drive a turbine pump through a discarded truck halfshaft, crown wheel and pinion. With the power in the wind being proportional to the cube of its speed, the output power/speed characteristics of the fixed-pitch blades was thereby matched to the load/speed characteristics of the pump.

This account has necessarily omitted to mention the vast number of smaller horizontal-axis windmills that have been invented or produced in all parts of the world. Yet, although only a few such small machines are still in commercial production, it seems that inquiries for them by prospective purchasers are once again on the increase.

Inventors have always been attracted by machines that rotate about a vertical axis and so avoid the need for orientation. The earliest machines of this type for which there is any reliable record were those used by the Persians for grinding corn. Several large sails were fitted to a vertical spindle, and the whole assembly was enclosed within a circular wall. Gaps in the wall admitted winds from different directions on to one side of the rotor and so forced it to rotate. There are many descriptions to be found in the literature of the host of similar vertical-axis machines that have subsequently been invented.

They are often known as 'solid-type' machines, so called because they interpose active rotor surfaces continuously across almost the entire area of the moving column of air. This is in contrast to propellor- type windmills, in which only a fraction of the swept area is occupied by the rotor surface. Lacroix used the word 'panemone' to describe an alternative solid-type machine. It features a series of fixed or movable vertically mounted blades, with provision for reducing the wind loading on those moving upwind.

Perhaps the best example of a vertical- axis machine is the Savonius rotor. It consists of two or more modified half cylinders with closed ends displaced sideways and arranged to rotate about a vertical axis. The open sides, which face in opposite direc-tions, catch the wind in a manner analogous to the cup anemometer. There have been many variants, but, because the whole of the swept area has to be covered by sheet metal, the constructional costs of these machines are inevitably large in relation to the output power extracted from wind. It can be shown that they have a maximum theore-tical efficiency of 0·33 (compared with 0·593 for a propellor- type of rotor).

In situations where there are scrap materials, such as old oil drums, available at a negligible cost, the simple Savonius rotor could make an economic contribution to small-scale water pumping or local electricity generation. A do-it-yourself design along these lines has been evolved by the Brace Research Institute for use in underdeveloped countries.

Finally, a novel vertical-axis machine was recently demonstrated by Raj Ranji and Peter South of the National Research Council of Canada (References 15-16). It uses two or three or more aerofoil-section blades bent into a catenary shape and fixed to hubs at both their upper and lower ends. In rotation under aerodynamic and centrifugal forces, the stresses in the blades are predominantly tensile. Preliminary tests show that high aerodynamic efficiencies comparable with the best propellor- type machines are readily attainable, but the design is not self starting. If further tests should confirm the early promise, the new design seems assured of a future.

This account has shown that most of the modern wind-power developments took place between 1940 and 1960. Rotary motion was almost exclusively the object of the initial energy conversion, although, it should be added, oscillatory designs have occasionally been produced. Research during this active period showed that a fixed pitch 3-bladed rotor was likely to give the lowest constructional costs consistent with minimising the problems arising from blade vibration. Provided that the wind rotor was mounted upwind of the tower, blade stresses could be kept acceptably low.

The design of small machines, up to, say, 10 kW, seems to be largely solved, but there remain some serious difficulties associated with the larger machines.

Scattered over the world, there is already a large volume of experience in the use of wind power. Because the power is not firm, it inevitably calls for a modified approach if it is to be used effectively. The energy available at any site can now be estimated with a fair degree of accuracy, provided that simple wind measurements can be made to establish the prevailing wind regime. Perhaps we are now once again poised at the beginning of a period in which renewed interest will bring about further developments in this nonpolluting source of energy.

References

1   PUTNAM, P.C.: 'Power from the wind' (Van Nostrand, 1948)

2   GOLDING, E.W.: 'The generation of electricity by windpower' (E. & F.N. Spon Ltd., 1955)

3   GOLDING, E.W., and STODHART, A.H.: 'The potentialities of wind power for electricity generation (with special reference to small-scale operation'). ERA Tech. Report W/T16, 1949

4   GOLDING, E.W., and STODHART, A.H.: 'The selection and characteristics of wind power sites'. ERA Tech. Report C/T108, 1952

5   VILLERS, D.E.: 'The testing of wind-driven generators operating in parallel with a network'. ERA Tech. Report C/T116, 1957

6   GIMPEL, G., and STODHART, A.H.: 'Windmills for electricity supply in remote areas'. ERA Tech. Report C/T120, 1958

7   WALKER, J.G.: 'The automatic operation of a medium-sized wind-driven generator running in isolation'. ERA Tech. Report C/T122, 1960

8   GOLDING, E.W.: 'The influence of aerodynamics in wind power development'. AGARD Report 401, 64 rue de Varenne, Paris VII, Sponsored by NATO

9   TAGG, J.R.: 'Wind data related to the generation of electricity by wind power'. ERA Tech. Report C/T115, 1957

10   JUUL, J.: 'Investigation of the possibilities of utilization of windpower'

11   JUUL, J.: 'Report of results obtained with the SEAS experimental wind-power generator', Elektroteknikeren, 1951, 47, pp.5, 12

12   JUUL, J.: 'Supplement to report on results obtained with the SEAS experimental wind-power Generator', ibid., 1952, 47, pp.65-67

13   VAN HEYS, J.W.: 'Wind und Windkraftanlagen' (Georg Siemens Verlagsbuchandlung, Berlin, 1947)

14   LEVY, N.: 'Current state of windpower research in the soviet union'. Tech. Report T56, ed. G.T. Ward. Brace Research Institute of McGill University, Canada, Sept. 1968

15   SOUTH, P., and RANJI, R.S.: 'Preliminary tests of a high speed vertical axis windmill model'. National Research Council of Canada Lab. Tech. Report LTR-LA-74, Ottawa, March 1971

16   SOUTH, P., and RANJI, R.S.: 'A wind tunnel investigation of a 14 ft.dia. vertical axis windmill'. National Research Council of Canada Lab. Tech. Report LTR-LA-105, Ottawa, Sept. 1972

Part 3

Energy demands

3.1   Geography and energy

Prof. E.M. Rawstron
Professor of Economic Geography, Queen Mary College, London, UK.

No policy should aim to achieve geographical parity between the pattern of energy consumption and the pattern of population distribution, because the need for energy will always vary from person to person and from activity to activity. But the massive geographical disparities that now exist deviate enormously from the requirements of economic and social efficiency, and thus urgently call for a policy that will quickly reduce them. While Los Angeles is polluted by the waste from using too much energy in a difficult natural environment, Calcutta is polluted by the lack of an energy supply sufficient to support a satisfactory standard of life.

1   Energy consumption

The consumption of energy (Fig.1) bears little quantitative relationship to the pattern of population (Fig.2). India (population 537 million) uses less than Italy (population 53·2 million). South and Central America together (population 274 million) use less than the United Kingdom (55·5 million). Canada's consumption (population 21·1 million) is 85% greater than that of the whole of the African continent (population 300 million).

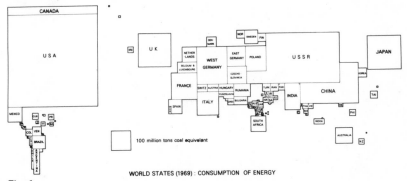

WORLD STATES (1969) : CONSUMPTION OF ENERGY

Fig. 1

The most striking features shown in Fig.1 are : the huge size of the USA compared with Latin America, and the huge size of Europe compared with Africa. India and China together are smaller than the USSR. Indonesia, which appears large in Fig.2, is far smaller than Australia in Fig.1 and not much bigger than New Zealand. Pakistan (before the secession of Bangladesh), with over 12 times the population of Greece, consumes only 1·17 times as much energy. Figs. 1 and 2 merit careful reading, together with the data in column 1 of Table 1.

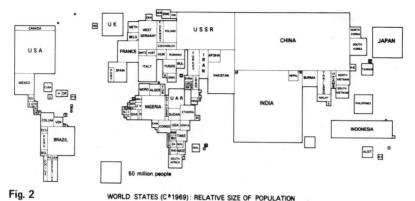

Fig. 2   WORLD STATES (C ᵃ 1969): RELATIVE SIZE OF POPULATION

Table 1

Energy consumption

	1969 kg per capita	1955-69 kg change per capita
WORLD	1804	603
Africa	291	*
Algeria	470	250
Egypt	221	−22
Ghana	155	37
Liberia	307	285
Libya	482	298
Morocco	195	34
South Africa	2746	359
S. Rhodesia	564	−
Tunisia	248	74
Zambia	509	300
North America	10585	3034
Canada	8794	3515
USA	10774	3006
Central America	1009	337
Colombia	591	176
Costa Rica	331	117
Cuba	1053	410
Guatemala	234	133
Haiti	31	−5
Honduras	230	104
Jamaica	1032	674
Mexico	1044	402

Table 1 (Continued)

	1969 kg per capita	1955-69 kg change per capita
Central America (Cont'd)		
Nicaragua	328	177
Panama	1348	962
Puerto Rico	3068	2234
Trinidad	4468	2665
South America	696	252
Argentina	1544	562
Bolivia	218	73
Brazil	481	192
Chile	1210	687
Ecuador	270	136
Guyana	896	373
Paraguay	131	71
Peru	623	349
Uraguay	914	259
Middle East	636	*
Cyprus	1252	807
Iran	562	404
Iraq	623	364
Israel	2154	1030
Jordan	308	194
Kuwait	12588	10914
Lebanon	689	213
Saudi Arabia	755	533
Turkey	461	237
Asia except Middle East	437	*
Burma	58	25
Sri Lanka	118	31
Hong Kong	931	524
India	193	79
Indonesia	98	−18
Japan	2828	2088
Sabah	365	*
Sarawak	273	*
West Malaysia	452	*
Pakistan (includes Bangladesh)	93	51
Philippines	261	135
Thailand	197	147

96

Table 1 (Continued)

	1969 kg per capita	1955-69 kg change per capita
Europe except Eastern Europe	3525	1209
Austria	2995	1134
Belgium	5429	1345
Denmark	5142	2646
Finland	3576	2392
France	3518	1352
Germany	4850	1593
Greece	1150	809
Ireland	2953	1833
Italy	2431	1524
Netherlands	4661	2213
Norway	4430	2050
Portugal	603	257
Spain	1354	642
Sweden	5768	2797
Switzerland	3172	1499
UK	5139	398
Yugoslavia	1243	544
Oceania	3901	1202
Australia	5200	1655
Fiji	407	182
New Zealand	2643	747
Planned Economies	*	*
Albania	608	452
Bulgaria	3491	2703
Czechoslovakia	6120	2234
GDR	5697	1819
Hungary	2888	980
Poland	4052	1431
Rumania	2628	1590
USSR	4199	1959
Asia, principally mainland China	505	346

Source: United Nations Statistical Yearbook

* Not available

In column 2 on Table 1, the per-capita change in energy consumption between 1955 and 1969 is shown. The areas with increases of more than twice the world mean include North America, much of Europe, Australia and New Zealand, and most of the communist countries. Notable exceptions are : the United Kingdom with only two-thirds of the average world amount (possibly thus helping to explain the poor growth rate of the British economy during the period but also reflecting an increase in thermal efficiency); Hungary with a substantially lower increase than the other countries of Eastern Europe, and South Africa. The increase in China, though less than the mean for the world as a whole, far exceeded that of India, and was substantially greater than that of South America. For the most part, therefore, the absolute increase per capita, between 1955 and 1969, was greater than that of the world as a whole in those areas that already used a far larger proportion of the available energy supply in 1955 than their population size would seem to warrant. The absolute gap between the well supplied and the poorly supplied areas has thus been widening.

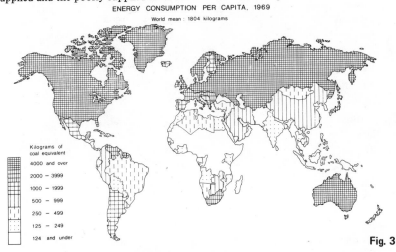

ENERGY CONSUMPTION PER CAPITA, 1969

World mean : 1804 kilograms

Kilograms of
coal equivalent

4000 and over

2000 — 3999

1000 — 1999

500 — 999

250 — 499

125 — 249

124 and under

Fig. 3

The top two grades of shading in Fig. 3 indicate the areas that use substantially more energy per capita than the world as a whole. Yet it is mostly these same areas that record both great absolute increases in energy consumption and, as Fig. 4 shows, low rates of increase in population.

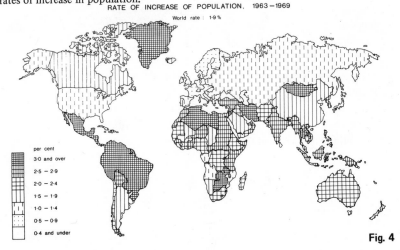

RATE OF INCREASE OF POPULATION, 1963 – 1969

World rate : 1·9 %

per cent

3·0 and over

2·5 — 2·9

2·0 — 2·4

1·5 — 1·9

1·0 — 1·4

0·5 — 0·9

0·4 and under

Fig. 4

Where rates of population growth are high, however, rates of energy consumption are usually low. The heavily shaded areas in Fig. 4 are mostly lightly shaded in Fig. 3, and, among the major so-called developing countries, it may be that China is the only one to show reasonably promising trends.

In sum, therefore, whether one views consumption of energy geographically at one point in time or as a trend through time, huge disparities are found to exist.

2   Reserves

There is as yet no worldwide shortage of energy resources. Fossil fuels are plentiful in the ground, and the affluent world has been so oblivious to any likelihood of future shortage that it has shifted its consumption in recent years away from the greater resource, coal (including lignite), towards the smaller but more convenient resources, petroleum and natural gas. The resource requiring a high input of labour has thus declined in exploitation relative to the resource requiring a low labour input. Considered jointly, however, there is no likelihood of a general shortage of fossil fuel during the remainder of this century, but local and regional shortages could arise.

The uneven spatial distribution of resources may contribute to these shortages, and may hinder the development of trends leading towards a more even pattern of consumption. Coal is plentiful in North America, the Soviet Union and Europe. China, too, is reasonably well supplied. The areas where coal is relatively scarce are south Asia, Africa (except South Africa) and Latin America. Few countries in the world are as well and conveniently endowed by nature as Britain is, with her close pattern of coalfields. Among major states, France and Italy are, in contrast, very poorly endowed, and, although Japan has considerable reserves on paper, their accessibility and quality are poor. If it became necessary in due course to turn back towards coal as the chief source of energy, certain parts of the world, notably, North America, the USSR, Poland and the United Kingdom, would be better able to make the change than many others. But success in the recovery of coal production in the future depends also on maintaining sizable and technically efficient coalmining industries during the present period of shift in consumption from coal to oil.

Reserves of petroleum and natural gas are deemed small in comparison with those of coal, and pessimistic forecasts have often been made about their likely duration. Large areas of the world still remain unexplored, however, and oilfields known to exist do not reveal the full extent of their reserves until several years after exploitation has begun. Thus, although one can be sure that oil and natural gas are jointly scarcer than coal, one can also be sure that the reserves of both are consistently underestimated.

The uneven spatial distribution of oil reserves is notable for the marked absence of major proven oilfields from the really densely peopled areas of the world. There are no major fields in China, India, Pakistan and Bangladesh. There are none in the populous north-eastern industrial belt of the USA. There are none in Europe except for the recently discovered field in the continental shelf of the North Sea. There are none in the populous western part of European Russia. Oil reserves and people do not seem to mix.

To supply the present pattern of demand, oil production normally involves transport over considerable distances from oilfield to market. Where this transport takes place within the bounds of one state, as in the USSR, long-term supply is more secure than where two or more states are involved. It follows therefore that, if the demand for oil continues to grow rapidly and tends to exceed supply, and if a substantial part of this increased demand arises in the developing world, as it will have to do if living standards there are to be raised, oil consumption will tend to contract on the regions in which reserves are located. India, China, Japan and substantial parts of Europe may find oil supplies difficult to obtain. Fortunately for these areas, a large proportion of the world's known reserves occur in the deserts of the Middle East and North Africa where export of oil is likely to exceed local demand almost until the reserves are exhausted.

Water power is neither plentiful nor always cheap compared with coal, but there are large untapped reserves, specially in the developing world, where some of the available sites should provide very cheap run-of-stream power. Tropical Africa is particularly well endowed, and there are many promising sites on the rivers flowing from the plateaux to the plains of South America. The large development already of hydroelectric sites in Japan would lead one to expect that similarly plentiful, but individually small, sites must exist in the island chains of the western Pacific and Indonesia. Large reserves await exploitation in the mountainous north of the Indian subcontinent.

Much of the potential of the affluent world has already been developed. But the contribution that hydroelectric power makes now, and may in future make, to the energy needs of that part of the world is comparatively small and must remain a minor source of supply save locally and regionally, as in Scandinavia and Canada.

The same cannot yet be said of the developing world, because its reserves are greater and because it consumes so little energy. Any hydroelectric development is likely therefore to be of major importance. Thus the installation of hydroelectric plants and transmission networks, even if they are used far below capacity for many years, would be a gesture, altruistic in the short term, on the part of the affluent world that could go far towards raising living standards in the tropics, and could prove highly beneficial commercially for the donors in the longer term, without either polluting the atmosphere or using up nonrenewable resources.

The future of atomic power seems likely to be a function of technological change, demand for energy and the competitive ability of other fuels, rather than a function of reserves of atomic fuel. If the demand for energy continues growing as quickly as present trends indicate, the incentive will increase for nuclear research to solve the problem of scarcity of uranium-235. Breeder reactors are already being developed; substitutes will be more thoroughly investigated, and research into nuclear fusion will be hastened.

An important fact to be noted about nuclear power is that supply of fuel places no restriction in peace time on the location of power plants, which do not therefore have to be sited either near their fuel supply or at a point to which large quantities of fuel can be cheaply transported. The nuclear plant, given a reasonable safety factor of distance and a site that fulfils certain technical needs, is thus a market-oriented source of energy. It is likely to be able 'very conveniently to satisfy the energy demands of much of the developing world, provided that the necessary capital can be found. Like hydroelectric schemes, therefore, aid given by the affluent world for the establishment of nuclear power stations in, for example, Bangladesh, India, Indonesia, Kenya, Pakistan and Peru to provide cheap, subsidised energy could give these lands the boost necessary for their economies to take off in pursuit of the developed economies of the affluent world. [There are many other considerations, highlighted in other papers in this book, that make a nuclear solution to this important problem undersirable — Editor]

3   Production

Consumption and production of energy used to coincide geographically. When the Black Dyke Mills were established at Queensbury in the middle of the 19th century to process wool for the worsted trade by means of steam-driven machinery, coal was produced from shallow pits in the neighbouring fields and taken by horse-drawn cart to the mill. The same space relationships were to be found, not only throughout the wool-textile area on the Coal Measures of the West Riding, but also in the Black Country, the Potteries, the Lancashire cotton district, and many other industrial areas of Britain. Coal was moved only very short distances before 1850, except by sea or inland waterway, and, for the most part, fuel had to be consumed very near to where it was produced. Output was determined therefore by the size of local demand, the most notable exception to this rule being the shipment of coal from the estuaries of the Tyne and Wear to London.

With the coming of railways to many parts of the world during the second half of the 19th century, the distances over which fuel was carried increased, but the amount transported more than 200 miles was small, and it is noteworthy that, even in 1913, the overseas sales of coal from Britain (the chief exporter at the time) amounted to only one-third (94 million tons including bunkers) of her total output.

While it should not be overlooked that sizable quantities of coal are transported great distances even today, it is oil that dominates international and even intranational trade in fuel. Unlike coalfields, oilfields have failed to attract much human settlement by local development of industry and commerce, and thus the pattern of production of fossil fuel no longer conforms as closely as it did, even 40 years ago, to the pattern of consumption. The major instances of discordance internationally relate to production in the Middle East, North Africa and the Caribbean, but intranational discordance is very considerable in both the USSR and the USA.

These spatial discordances have developed to meet the enormous expansion of energy consumption that has taken place in the affluent world since the war, and specially since 1956. At that time OECD and other authorities were predicting a shortage of fuel for the foreseeable future, but this prediction was refuted almost immediately by the glut that began in 1958 and continued through the next decade. Now fears of imminent shortage in particular parts of the affluent world, notably in the USA, are once more in vogue, and, to allay them, further spatial discordance between consumption and production will be inevitable.

4   Energy and wealth

A comparison of Fig. 5 (relative sizes of gross national product) and Fig. 1 (consumption of energy) shows that the two patterns are very similar. It is safe to assume that there is a close causal relationship between them and that, generally speaking, a high standard of living demands a high consumption of energy. It follows that, if the production of energy continues to be aimed at satisfying the increasing demands of affluent states rather than the latent needs of poor ones, the distortions apparent on Figs. 1 and 5 will increase rather than diminish, and, as others have observed analogously in the general economic context, both the energy gap and the wealth gap will widen between rich and poor nations.

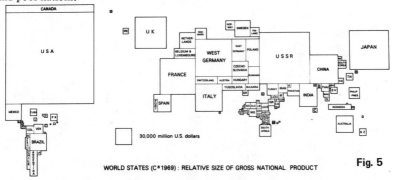

WORLD STATES (Ca 1969): RELATIVE SIZE OF GROSS NATIONAL PRODUCT   **Fig. 5**

101

The so-called energy crisis is not then mainly a matter either of shortage for the time-being or of environmental pollution through locally excessive and inefficient combustion. It is a matter of maldistribution and misuse. The first aim of any international policy for energy should be to survey needs with a view to reducing spatial disparities of consumption per capita, the implication being that output must be massively increased before it can be stabilised at a level conformable with available reserves, the establishment of a satisfactory environmental control, and the rate of population change. Extensive aid for the development of hydroelectric and nuclear power is clearly going to be required without delay. The second aim should be to raise the efficiencies of energy processing, for example, in coal-fired and nuclear power stations, and of energy use, for example, in general heating and transport. The increasing use of waste heat and the application of the right kind of power for the job are two obvious requirements.

Plentiful energy is essential for human advancement and general welfare. Yet much of the world uses too little, while a little of the world uses too much. Spatial maldistribution is the crux not only of the geography of energy but also of the problem of relating energy to human welfare in the decades to come.

5. References

1   Statistical Year books of the United Nations

2   'Indicators of market size', Business Internat.
    December 4th, 11th and 18th and January 1st

3   The Times Concise Atlas, 1972, pp 20-21 and 24-25.

3.2 Energy and nitrogen fixation

B.O. Gillberg
Department of MicroBiology, University of Uppsala, Sweden.

Agricultural products result from direct energy conversion from the sun, whereas industrial products are manufactured using energy derived from fuels. Economists demand that we increase productivity, and, to do so, we must put in more and more energy. This is possible in industry, but agriculture is governed by biological laws, which exert limits on the speed of processes.

Nevertheless, farmers do put in energy — not only by replacing muscle power, such as that of the horse, with mechanical power from tractors — but also by the use of chemicals. Pesticides and nitrogenous fertilisers require vast quantities of energy for their production. This amounts only to about 1% of the total energy production of the industrialised world. Should a lesser industrialised country, for instance Egypt, however be involved in producing its own fertiliser needs, this would amount to 14% of the total electricity output of that country.

The trend is for the developed countries to depend more and more on increased usage of fertilisers; e.g. Sweden between 1935 and 1970 increased its production of fertilisers to 210 000 from 19 000 tons (nitrogen), but the returns, in terms of crop production, have not continued to increase. What is more serious is that we are exporting this policy to the Third World. They cannot afford the energy necessary to produce fertilisers, in anything like the quantities that would be needed, if they were to use the high loading levels (up to 600 kg per hectare) now reached. What is more, energy is needed for export and transportation of these fertilisers, and transportation is itself a problem on many of the roads of the developing countries. They thus become more dependent on the industrial nations for their supplies of fertilisers.

It is a tragedy, considering the lack both of protein and of energy in the world today, that the policy adopted to increase protein production should be that of the highly energy-dependent method of increased fertiliser treatment.

The chronic worldwide shortage of dietary protein reflects a fundamental deficiency in available nitrogen for plant protein synthesis. There are very large quantities of nitrogen in the atmosphere, but the plants cannot use gaseous nitrogen directly. However, both freeliving and symbiotic micro-organisms, in the soils, are capable of transforming gaseous nitrogen to socalled fixed nitrogen (nitrate etc) that can be used by the plants for protein synthesis. This process is called biological nitrogen fixation. It has been estimated that around 90 million tons of nitrogen are biologically fixed in the soils every year. This is roughly 90% of the total amount of nitrogen made available in the soils (Reference 1). The fertiliser manufacturers produce approximately 10% of the nitrogen needed to grow the world crops, the rest is of biological origin (Reference 2), 70% of the total world chemical-fertiliser plants are located in Western Europe and the USA, where 18% of the world population is living (Reference 1).

103

It is, of course, possible to increase the agricultural productivity in nitrogen-poor soils by using nitrogen fertilisers. However, as a result of the intensive use of fertilisers, an appreciable part of it leaches out from the soils as nitrate and enters the surface waters, where it becomes a serious pollutant that may even affect human health. Hazardous amounts of nitrate have been found in drinking water in the USA and in food in certain countries owing to the heavy use of fertilisers (References 3-5).

Against this background, it is obvious that, for a long time, probably for ever, man must depend on biological nitrogen fixation – certainly in the Third World, where the farmers cannot afford fertilisers, and where lack of roads inhibits the transport of large quantities of fertilisers.

The best known and, from an agricultural point of view, the most important system of biological nitrogen fixation undoubtedly is the Rhizobium-legume system. Bacteria belonging to the genus Rhizobium fix atmospheric nitrogen in symbiosis with leguminous plants. The nitrogen fixation takes place in nodules induced on the plant roots by the bacteria, and Rhizobium is often called the nodule bacterium. Rhizobium is capable of fixing nitrogen only in symbiosis with legumes, not as freeliving organisms. Only two groups of symbiotic nitrogen-fixing systems are known: the Rhizobium-legume system and the association between a micro-organism, possibly an actinomycete, and certain trees and bushes like Alnus and Casuarina.

Field experiments in New Zealand have shown that Rhizobium, in association with clover, can fix the equivalent of about 500 kg/ha of nitrogenous fertiliser, such as sulphate of ammonia (Reference 2). Under temperate climatic conditions of Western Europe, Rhizobium and clover or lucerne may fix around 300 kg of nitrogen per hectare annually (Reference 6).

The mechanism of symbiotic nitrogen fixation is rather unclear. However, in recent years, nitrogen fixation has been obtained in cellfree systems, using enzymes from nitrogen-fixing organisms (References 7-9). The mechanism of nodule formation is far from understood, and further studies in this field might be very fruitful. When, if ever, we understand the chemistry of symbiotic nitrogen fixation in legumes, we might be able to establish symbiotic nitrogen fixation in major crops such as wheat or corn.

Rhizobium is classified into 16 different groups according to its ability to form nodules on the legume hosts (Reference 10). Each species forms nodules only on a certain number of legume genera. A strain causing nodulation on clover thus belongs to the clover group; it does not form nodules on other legume hosts. These specific groups are often called cross-inoculation groups.

The capacity of Rhizobium to fix nitrogen in symbiosis with the plants varies greatly. The capacity to fix nitrogen is known as effectiveness. Certain strains are highly effective, and others form nodules but do not fix nitrogen. Those that do not fix nitrogen are called ineffective.

In Western Europe, the USA, Australia and New Zealand, Rhizobium has been used since the beginning of this century to increase the production of legumes. Handling of the bacteria is quite simple. Inoculants (bacteria cultures) are often distributed in small cans or bottles to the farmers. The content is mixed with the seed immediately before sowing. The cost of treatment of seed for 1 ha is about $ 2-4 (Swedish prices), whereas nitrogenous fertilisers for the same area today cost about $60. The weight of the inoculant for 1 ha is ca. 0·5 kg, whereas farmers today might apply as much as 500 kg/ha of nitrogenous fertilisers. Inoculation of legume seed with Rhizobium is very common today in the industrialised countries where large quantities of legumes are grown. In the USA, more than 75% of the soybean seed is inoculated. The figures are about the same for other important legumes such as Alfalfa (Reference 11).

The soils of the 'undeveloped' protein-deficient areas of the world are very often poor in nitrogen, so that it is not possible to grow wheat, corn, potatoes etc., without some kind of nitrogenous fertilisers. However, protein-rich leguminous plants such as soybean and chickpea can be grown without costly fertilisers, provided that the seed is inoculated with Rhizobium. This is not the practice in the Third World today. Growing more legumes and inoculating them with Rhizobium might be one way to increase the protein production in the starving parts of the world. This has also been recommended by the UN and the FAO (References 12-13). The following is said in an FAO report: 'The production of legumes should be increased in a large part of the tropics and subtropics, and will lead to a real improvement in nutrition'(Reference 11). This recommendation is obviously valid, when it is realised that the meal left after extracting soybeans for oil contains about 40-50% of good-quality protein that can be used directly for human consumption (Reference 12). Also the protein nutritional value of soybean is among the highest of all proteins of vegetable origin, and soy bean proteins, compared with other plant proteins, are rich in lysine and therefore useful as a supplements to cereals (References 14-15). Legumes like the soybean are also very efficient protein producers. One acre of edible soybeans can produce enough protein to supply an active adult man's protein requirement for about 2200 days, whereas (Fig.1) one acre of wheat (white wheat flour) will supply the protein requirement for about 500 days and one acre of beef cattle will cover the requirement for only 77 days (Reference 16). Growing more legumes is obviously a way to increase the global production of proteins for direct human consumption.

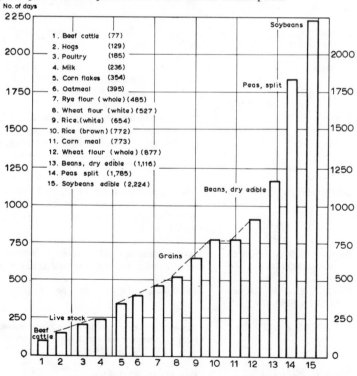

Fig. 1

Number of days that an active adult's protein requirement can be obtained from one acre-year, depending on the crop or livestock that is kept on it (from Potential of Animal–, Fish–, and certain Plant - Proteins. I Dairy Science 52, No. 3, 409(1969))

Legumes are also very useful when it comes to increasing the soil fertility. In the poor South Australian soils, for example, there has been a great soil improvement after growing subterranean clover. After a pasture period of 25 years, there is a rise of the nitrogen content of the upper 10 cm soils from 0·06% to 0·18% and, of course, also a building up of organic matter that is of importance for soil fertility (References 17-19). An increased productivity in tropical grassland areas after introduction of legumes has also been reported (References 20-22). Legumes have also been found to be useful in plantations (tea, rubber, coffee etc.) for erosion control, for preventing leaching of nutrients, and for creating a good soil structure (Reference 6). Rubber trees grow bigger and come earlier into production with a ground cover of legumes instead of the natural ground flora.

Growing more legumes and inoculating the seed with efficient strains of Rhizobium seems to be one way to increase the global protein production. However, it is not a question merely of inoculating the seed and sowing it, because Rhizobium strains that are efficient nitrogen fixers in one environment do not always function equally in another (References 6, 23-24). Soil conditions such as pH, soil temperature, pesticide residues, antagonistic micro-organisms etc. are of great importance in an effective symbiotic process. This means that not only the plant but also the bacterium must be selected or bred for the environment in which it is to be used, to get an optimal nitrogen fixation.

References

1	STEWART, W.D.P.	'Nitrogen-fixing plants', Science, 1967, **158**, pp. 1426-1432
2	WALKER, T.W.	'Legume inoculation in New Zealand'. New Zealand Department of Scientific & Industrial Research Information Series 58, 5 (1966)
3	COMMONER, B.	'Threats to the integrity of the nitrogen cycle: nitrogen compounds in soil, water, atmosphere and precipitation'. Presented at the annual meeting of the American Association for the Advancement of Science, Dallas, Texas, 1969
4	COMMONER, B.	'The Origins of the environmental crisis'. Keynote address before the Council of Europe, 2nd symposium of Members of Parliament specialists in public health, Stockholm, 1971
5	KOHL, D.H., SHEARER, G.B. and COMMONER, B.	'Isotropic analysis of the movement of fertilizer nitrogen into surface water', Science (to be published)
6	MULDER, E.G., LIE, T.A, and WOLDENDORP, J.W.	'Soil biology' (UNESCO, 1969), pp. 171-172
7	BERGENSEN, F.J.	'Some properties of nitrogen-fixing breis prepared from soybean root nodules', Biochem. Biophys. Acta, **130**, 1967, pp. 304-312
8	KOCH, B., EVANS, H.J. and RUSSEL, S.A.	Plant Physiol, **42**, 1967, p.466
9	HARDY, R.W.F., BURNS, R.C., and PARSHALL, G.W.	'The biochemistry of N_2 fixation', Advances in Chemistry Series 100, 1971, pp.219-247

10	FRED, E.B., BALDWIN, I.L., and McCOY, E.	'Root nodule bacteria and leguminous plants' (University of Wisconsin Press, Madison, 1932)
11	WEBER, D., and BURTON, J.C.	Private communication
12		'International action to avert the impending protein crisis'. UN publication E68 XIII,2, 1968
13		'Legumes in human nutrition'. FAO Nutritional Studies 19, 1964, pp.88-99
14	MILNER, M.	'General outlook for seed protein concentrates'. 'World protein resources'. Advances in Chemistry Series 57, 1966
15		Agricultural Research Service, US Department of Agriculture, Proceedings of the conference on soybean proteins in human foods, Peoria, III, 1966
16	WILCKE, H.L.	'Potential of animal-, Fish- and certain plant-proteins', J. Dairy Sci., 1969, 52, (3), p.409
17	DONALD, C.M.	'The progress of Australian agriculture and the role of pastures in environmental change', Aust. J. Sci., 1964-65, 27, pp. 187-198
18	DONALD, C.M., and WILLIAMS, C.H.	'Fertility and productivity of podozolic soils as influenced by subterranean clover (Trifolium subterranean L.) and superphosphate'. Aust. J. Agr. Res., 1954, 5, pp.664-687
19	RUSSEL, J.S.	'Soil fertility changes in the long-term experimental plots at Kybybolite, South Australia', Aust. J. Res. 1960, 11, pp.902-926
20	ANON	'Some concepts and methods in subtropical pasture research'. Commonwealth Bureau of Pastures & Field Crops, Hurley, Berks., England, Bulletin 47, 1964
21	McILROY, R.J.	'Grassland improvement and utilization in Nigeria', Outlook on Agr. 1962, 3, pp.174-179
22	SMITH, C.A.	'Tropical grass/legume pastures in Northern Rhodesia', J. Agr. Sci., 1962, 59, pp.111-118
23	GILLBERG, B.O.	'On the effects of some pesticides on Rhizobium and isolation of pesticide-resistant mutants', Arch. Mikrobiol., 1971, 75, pp. 203-208
24	FLETCHER, W.W. and ALCORN, J.W.S.	'The effect of translated herbicides on Rhizobia and the nodulation of legumes' Hallsworth, E.G. (Ed), 'Nutrition of the legumes' 1958

3.3   Energy and the car

G. Leach
Freelance journalist

I should like to dedicate this essay to the majority of mankind which has no hope at all — and, I sincerely hope, no ambition — of owning an average American car. This paragon of ecological virtue has an astonishing capacity for swallowing non-renewable resources. Quite apart from its contribution to the death rate, air pollution, noise and the blighting of city and countryside, the typical US car spends much of its time transporting a mere one to 1·5 people, four to five empty seats and 1·6 tons of metal and other materials by consuming exactly two tons of gasoline each year — a figure that may not seem all that high until one realises that it is equivalent to a **continuous** 3kW, or a 3-bar electric fire left on day and night, 365 days a year for the 10 — 11 years that the average car lasts. What is more, there are now nearly 100 million of these spendthrift objects, or just under one for every two Americans.

Until recently, this kind of energy-consumption figure did not seem too important in the whole debate on the environmental impact of automobiles: accidents, noise, pollution and the blighting effects of mass automobility dominated the discussions. Now, as the spectre of oil shortages hangs over the industrial world, and particularly over the USA, the role of the car and of other forms of transport in consuming oil is beginning to be seen as an important factor.

What I want to do today is to demonstrate that it not merely is important but could be decisive; that, if present trends in car ownership and fuel consumption continue, road transport led by the car could squeeze world oil resources very tightly indeed by the 1980s. As a result, energy considerations are almost certain to become a prime factor in the design and planning of cars and of transport systems — a factor just as important as the present concern with, say, pollution. Further, I would also like to show that reducing energy consumption by the car will not be at all difficult: as my opening remarks about American cars suggest, most cars as conceived and used at present are extremely extravagant devices for burning fuels to shift people and goods about. It is not too difficult to imagine ways of conceiving and using them better from an energy-consumption point of view.

Let me start then with a few facts about the car's use of energy, culled from an extensive study I have done for OECD on the energy and metal demands of the world's fleet of cars and other road vehicles (Reference 1).

Table 1 gives a breakdown of the energy consumption by transport and various forms of transport in the USA and OECD Europe [roughly speaking, Western Europe (Reference 2)] in 1960 and 1969. Since the USA in 1970 contained 48% of the world's fleet of 187 million cars and OECD Europe contained a further 34% (Canada and Japan bring the total for the Western industrial world to 91%), the Table represents the vast majority of the world's cars.

Table 1

Transport's share of net energy consumption: USA & OECD Europe

	USA		OECD	Europe
	1960	1969	1960	1969
	(10^9 kWh equivalent)			
Total net energy consumption	8865	14747	5594	8876
Consumption all transport:	2815	4111	898	1447
as % of total	31·8	27·9	16·1	16·3
Consumption road transport:	2369	3356	535	1240
as % of total	26·7	22·8	9·6	14·0
as % of all transport	84·2	81·6	59·6	85·7
Consumption by passenger car:				
as % of total		16·1		9·4
as % of road transport		70·7		67·1
Consumption by air transport:	302	596	55	138
as % of total	3·4	4·0	1·0	1·6
as % of all transport	10·7	14·5	6·1	9·5

Source: OECD energy statistics 1955-69, pp. 188-205 and 111-129.

Note: All data from source given except those for passenger car, which come from various sources

'Net consumption' is total useful energy consumed; i.e.
excluding energy losses in electricity generation and fuels used for non-energy purposes.

One of the most striking points made by the Table is that the car really does dominate in terms of energy consumption. In both Europe and the USA in 1969, the car accounted for close to 70% of all energy used by road transport, and 58% of all transport energy. In contrast, air transport, which like the car and lorry also relies almost totally on vulnerable petroleum fuels, took only a quarter of the car's share of energy in the USA and one-fifth of the car's share in Europe.

One must emphasise that these figures are only for propulsion fuels. They do not include the many indirect sources of energy consumption, including the energy for mining and processing the materials that go into vehicles, the energy costs of manufacture and assembly, of road building and maintenance, and of the vast sales, advertising, garage and fuel-distribution networks. Several attempts to total these indirect fuel costs for the USA have produced answers in the region of 40% or so of net energy input for all transport, compared with the 28% shown in Table 1; and of indirect costs for the car amounting to nearly as much as the direct 'propulsion only' costs. A very careful analysis by Prof. Berry of Chicago University has found that the energy costs of manufacturing a car — including such nth-order costs as the energy required to transport iron and aluminium from mines and smelting plants to the car factory — amount to just under three tons of oil equivalent, or about the same as the average car uses in direct 'propulsion fuel' in 18 months (Reference 3).

But one does not have to include these indirect costs to demonstrate that the car and other road vehicles may be hastening an impending energy crisis, provided that current trends persist.

The two most important trends are clearly fuel consumption per car and the growth of car numbers. Taking the first, one finds that, on average, fuel consumption per vehicle is hardly changing. Nor is it expected to change much in future. Annual mileage driven increased only 3% in the USA and only 1% in W. Europe during the 1960s – a period in which the car population rose 45% in the USA and nearly trebled in W. Europe (from 22·6 million to 64·3 million vehicles). Specific consumption (miles per gallon) has not changed much either, nor is it expected to over the medium term. The effect of extra pollution-control equipment and of deleading petrol is widely expected to raise the fuel consumption of vehicles, but it is equally widely expected that, during the 1970s, these increases will be offset by improvements in fuel performance through better engine design and so forth. As a result, most oil companies are forecasting only the slightest of rises in fuel consumption per car by 1980. One leading British company reckons on a 3% rise, for France, Italy, Germany and the UK combined, by 1980. Another forecasts a 5% rise for these countries, but a 4% fall in the rest of the world, with a consequent rise for the whole world of only 2·5%.

If these predictions are valid, it looks as though the growth of car ownership, and hence of total vehicle numbers, is far and away the most important factor in influencing present predictions of future energy consumption by the car.

Forecasts of the growth of the car are in fact remarkably consistent – at least when governments or government agencies are doing the forecasting. A consensus view of these forecasts – compiled by OECD from 'official' sources (Reference 4) – is given in Figs. 1 and 2, though the extrapolations beyond 1985 are my own. Later, I will ask whether these forecasts of massive growth in vehicle populations can possibly be valid in view of the mounting pressures against the car in all highly motorised societies, but, for the time being, one might take them as valid. After all, they are the projections on which governments are now acting; so it seems perfectly legitimate to use them as a basis for forecasting what such massive growths will do to the future of world oil supplies.

When one does this, one arrives at some rather remarkable conclusions. But, first, what are present car populations doing to world energy and oil resources?

One of the findings of my study for OECD is that, whereas the average car in the USA consumes two tons of gasoline per year, in the rest of the world, the average figure is very close to one ton, and, for all commercial vehicles (i.e. lorries and buses), it is close to four tons per year. Combining these figures with numbers of vehicles, one finds:

(a)   In 1970 the world's cars took 12% of global oil consumption by weight: 277 million tons out of 2288 million tons.

(b)   Other road vehicles took nearly the same again, or 9% of gross oil consumption.

(c)   All road vehicles therefore took 21% of global oil. However, the US car fleet, with 48% of the cars, accounted for 65% of this total oil take.

Until recently, the world's oil industry did not feel too much strained by these proportions. It did not have any serious trouble in splitting each barrel of crude at the refineries into different types of fuel for different users – transport, power stations, industry, domestic heating, and so on. In Europe, it still has little trouble. Road transport causes it few headaches, because, first, only 17 – 18% of each barrel of crude goes into road-vehicle fuels; and, secondly, this proportion is actually declining slightly. In other words, other users of oil are increasing their demands more rapidly than motorists and commercial transport concerns.

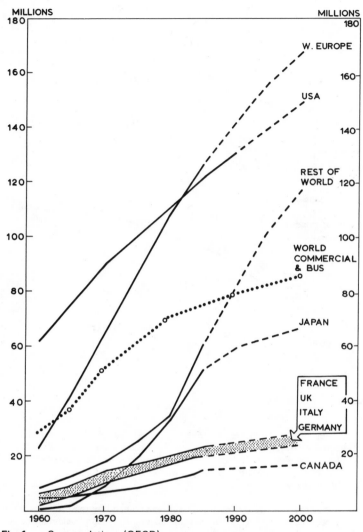

Fig. 1 Car populations (OECD)

This was also true in the USA until recently, though there the split of the barrel is very different. Instead of 17 − 18% of the crude going to road vehicles, very close to 50% does. In other words, road vehicles take one-half of all oil, by weight. For technical reasons, this is getting fairly close to the limits − one cannot get much more gasoline and diesel out of average crude than that without going to fairly extreme refining methods. But in one summer, these limits were strained to the utmost by an unpredicted 7% rise in demand for motor spirit, apparently because of the effects of de-leading gasoline, and the worsening fuel performance owing to anti-pollution devices in cars. Coupled with a shortfall in refining capacity, the result was widespread gasoline shortages in many States.

Will such shortages become a permanent and general rule, rather than a possibly temporary affliction for one country, the USA?

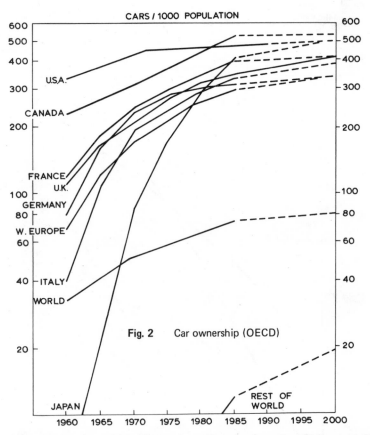

CARS / 1000 POPULATION

Fig. 2   Car ownership (OECD)

600
500
400
300
200

100
80

60

40

20

1960 1965 1970 1975 1980 1985 1990 1995 2000

U.S.A.
CANADA
FRANCE
U.K.
GERMANY
W. EUROPE
ITALY
WORLD
JAPAN
REST OF WORLD

Fig. 3 suggests that they might. The two humped curves show the widely accepted forecasts of how global oil production will peak and then decline, given the highest and lowest estimates of the total quantity of oil that will be recovered from the Earth. Superimposed on these are two curves, the same in each Figure, marked A and B which require some explanation.

What they measure are the 'required' crude-oil demand by road vehicles, given the assumptions used in this paper: i.e. that consumption per vehicle remains static and that the number of vehicles grows as shown in Figs. 1 and 2. The curve A further assumes that refinery outputs will remain as now in the 'split of the barrel' between road vehicle fuels and all other fuels. In other words, in the USA, 50% of crude oil will continue going to cars and commercial vehicles, with the consequence that, to provide one ton of fuel for such a vehicle, the refinery must take in two tons. These two tons are the 'required' crude-oil demand for one ton of motor-vehicle fuel. Similarly, in Europe and the rest of the world, the 'required' crude-oil demand for one ton of motor-vehicle fuel is five tons, since the refinery split is such that, for each five tons of crude arriving at the refinery, only one ton (approximately) goes to road vehicles, the other four tons going to other oil-consuming sectors of the economy. The lower curve B is based on the same overall reasoning, but with the important difference that a maximum effort is made to channel petroleum fuels to road vehicles: in the USA, the refinery split is assumed to remain as now, but, in the rest of the world, there is a steady increase between 1970 and the year 2000 in the proportion of oil that is channelled to road vehicles so that, in the year 2000, the refinery split is the same as it now is in the USA.

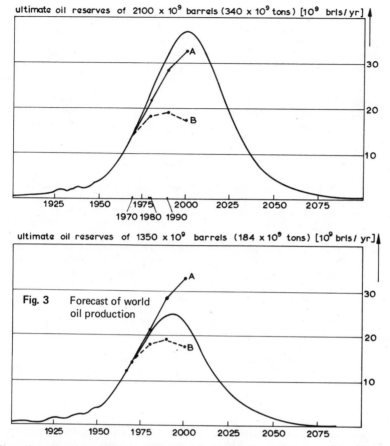

ultimate oil reserves of 2100 x 10⁹ barrels (340 x 10⁹ tons) [10⁹ brls/ yr]

ultimate oil reserves of 1350 x 10⁹ barrels (184 x 10⁹ tons) [10⁹ brls/ yr]

Fig. 3 Forecast of world oil production

Given these broad assumptions, what do we find? Clearly, if the highest estimate of oil reserves are correct, and the oil industry can find this much oil without causing intolerable pollution, beggaring itself, or causing a few wars between the haves and the have-nots, there is little to worry about as far as road transport's demands for oil go in this century.

But if the lower estimate is correct – and if present trends continue – the car and other road vehicles are clearly going to precipitate a fairly major oil 'crisis' around the late 1980s. Another way of looking at this is to say that, since this is a hypothetical doomsday projection – and cannot, in fact happen – the assumptions must be wrong. In other words, car ownership cannot grow as official projections predict, and/or the fuel consumption per vehicle will have to be reduced. The system will have to change.

How is it going to change? There seem to be many possible responses. First, I do not believe that the projections of car ownership quoted earlier are true. These projections are based on little more than the expectation that the rest of the developed world will, and wants to, catch up with present American levels of car ownership. America is already deeply concerned about all the effects of cars on its environment, specially cities, and is trying vigorously to reduce the dominance of the car. If this is true of America, it is doubly true of Europe, where land for roads is relatively more expensive, there are much stronger traditions of public transport and much easier conditions for public transport – including smaller and more compact cities and higher population densities – and a deeper concern for preserving ancient city centres from being choked by the car.

113

Secondly, it will not be difficult to reduce fuel consumption by the cars that we do have. At least, it will not be difficult if two conditions are fulfilled: if government and public opinion demand it, and if the motorist is sensitive to rising costs of oil and therefore to the real cost of running a car. For consider what might happen if he is completely insensitive. The car and road transport would require an accelerated exploitation of oil reserves. As pressure on oil builds up, prices will rise to the point where the oil industry will resort to higher-cost techniques of production — first, the expansion of offshore oil, tertiary recovery methods (steam injection, flooding wells with solvents), then the exploitation of tar sands, oil shales and the conversion of coal — all of them resources for oil that are very abundant but either remote or expensive or both.

In this situation — which, most agree, is the situation we are beginning to enter — the car as we know it now could flourish and grow quite a long time beyond the end of the century. But, if this does happen, the car would be a major part of a social philosophy that encourages the depletion of an irreplaceable asset, which encourages the relatively affluent to have what they want — private transport — and, by doing so, denies a precious resource to other inhabitants of the Earth and to future generations.

I hope this is not the path we will follow. I hope that either the motorist will be acutely sensitive to the rise in real costs of motoring due to rising costs of oil or that the community will decide on commonsense or moral grounds, the scenario I have just outlined is a nightmare, and decide instead to conserve its oil reserves by cutting down the intake by the car and road transport generally.

And this would be so easy. Even if car ownership grew as shown earlier, it would only take an annual fall of 4% in consumption per vehicle — for world oil consumption by road transport in year 2000 to be no higher than today. Is this a credible figure? I am sure it is. Consider some of the cuts that could be made:

(a) Make all US cars adopt the fuel economy of Europe. This would cut US annual oil consumption by 89 million tons or 17%, and world oil consumption by the car by one-third.

(b) General Motors has calculated that one could build a car with perfectly respectable performance — a range of 60-100 miles, top speed of 50-60 mile/h, weight 1 ton — with an engine of only 7·8 hp — or about one-eigth the present European average. If such a vehicle were universal today, world oil consumption by the car — all other factors being equal — would be only 24 million tons or so. In other words, just under one-tenth the present figure.

(c) Then, of course, we could shift the car away from oil. Electric cars do not help much here, as long as power comes through the very inefficient chain of fossil-fuelled or nuclear stations. Either way — internal-combustion engine or electric car — you end up with a net efficiency from the original fuel to motive power at the wheels of around 15% to 20%.

Without going into details, it does look as if electric cars en masse are not much of a solution (a) until we get safe and cheap nuclear power, and (b) until battery development has gone very much further, preferably towards using materials that are not resource-critical, such as the iron-air or sodium-sulphur battery.

But there are other nonoil energy sources for cars. Liquified natural gas is one. Identified reserves in the USA alone are estimated at 73 billion tons oil equivalent, or roughly 40% of the low estimate of ultimate world oil reserves. The total resource of US natural gas more than doubles this. So natural-gas cars would stretch out the world's oil considerably, though they would only postpone the problem.

Much more significant is the methane car, or bus or goods truck. I will not argue the very important merits of home-generated methane in the political philosophy of alternative or people's technology. I will just say that, if the present technocratic-economic system rationalised itself, ordinary garbage could generate very large quantities of energy. One estimate is that all US city garbage could provide methane equivalent to half US oil consumption, i.e. enough to run all US road vehicles.

Equally promising is high-temperature pyrolysis of organic wastes to produce liquid fuel. For example, Garrett Research & Development, a subsidiary of Occidental Petroleum, has a 4 ton/day pilot plant in California that shreds city garbage and gets 40% by weight yield of liquid fuel with heating value about half that of gasoline. The fuel is suitable for industrial boilers or as a feedstock to refineries for producing higher-grade fuels, such as gasoline.

Another way of reducing the car's demand for fuel is, simply, to reduce the use of cars! One could use cars much more efficiently by increasing occupancy, for example by car-sharing schemes. One could switch from cars to other road vehicles, whether standard types of public transport, such as the bus, or any of the plethora of advanced schemes now being developed, such as the small pickup bus or dial-a-bus. One could switch from the roads altogether towards traditional rail and metro commuter services, or, again, to many of the new advanced technology schemes now going into service or on the drawing board. Finally, of course, one could redesign cities and work patterns, so that much of the quite unnecessary toing and froing that goes on now is cut out — a Utopian concept.

I will not elaborate on the many possibilities of this kind for reducing the use of cars — a reduction, one must add, that has many other motives behind it than reducing oil consumption. But it is worth considering what effects such reductions would have on energy consumption, by looking at Figs. 4 and 5. These show that, with the exception only of the luxury Pullman train, the typical automobile with its one to two occupants is by far the most inefficient form of land transport in terms of energy consumption per passenger mile that has ever been invented. Aircraft exceed the car in inefficiency, often by wide margins, but then they do provide the important advantage of very much higher speeds.

These figures from Prof. Richard Rice, of the Carnegie-Mellon University, more or less tell their own story: the large car with a single occupant is a most energy-extravagant form of transport, the small car with a full load is not (it does as well or better than most forms of public transport), but there is nothing quite so good in an energy-hungry world as the bicycle. Simple solutions are often the best.

The real question highlighted by the car's use of energy is not so much whether the world will run short of oil because of the growth of motor vehicles, nor whether technical developments can avert such a crisis. The real question is how governments and communities generally will act in this situation.

If they continue to permit the private motorist almost unfettered freedom to own and to use cars with extravagent fuel consumptions — for example, by cushioning the shock of higher oil prices at the refinery through a reduction of gasoline taxes — the growing demands of motoring are almost certain to accelerate the exploitation of the world's dwindling oil reserves and thus to raise prices still further. This policy would allow the rich to continue enjoying the luxury of high-performance private transport, but it would also impose great strains and costs on the world community in the long term. It would merely postpone the problem of depleting oil supplies. It would raise oil prices — perhaps very sharply — and so force the abandonment or postponement of many other opportunities for development, thus leaving many urgent needs unfulfilled. All nations would be affected, but those that can least afford it would be hardest hit.

Ratio Scale

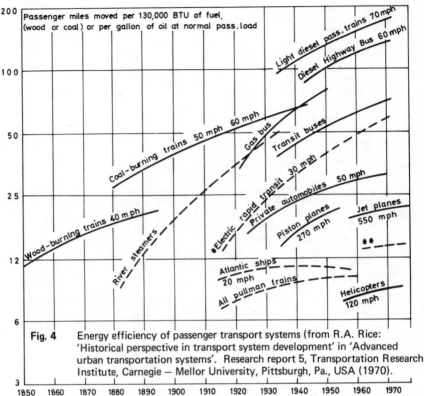

Fig. 4 Energy efficiency of passenger transport systems (from R.A. Rice: 'Historical perspective in transport system development' in 'Advanced urban transportation systems'. Research report 5, Transportation Research Institute, Carnegie — Mellor University, Pittsburgh, Pa., USA (1970).

A more rational (and moral) approach would be to recognise, at the earliest possible moment, that energy consumption should be a major factor in the design of future vehicles and transport policies, and that, consequently, every attempt should be made to reduce the very heavy energy demands — particularly oil demands — of the private car (and of other road vehicles). Many approaches should be considered, but action on four points would go a long way towards reducing both the social and energy costs of car ownership:

(a) the encouragement of public road and rail transport; of taxi or car-sharing schemes that increase vehicle occupancy; and of foot or bicycle traffic in cities

(b) the encouragement of smaller cars with lower gasoline consumption (and pollution characteristics), perhaps by differential taxation against larger engines

(c) the encouragement of research and development of batteries for electric cars and on the conversion of organic wastes to alternative transport fuels

(d) the encouragement of rail, pipelines and perhaps waterways, rather than of road vehicles, for freight transport.

116

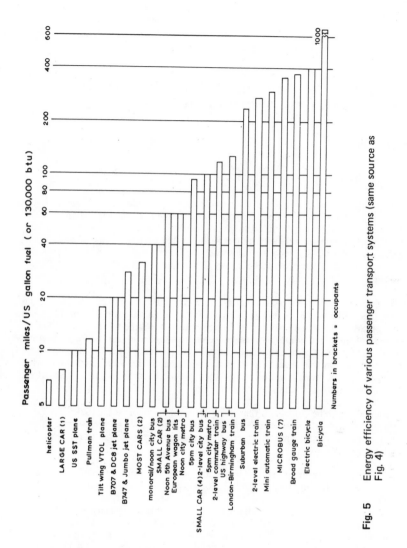

Fig. 5 Energy efficiency of various passenger transport systems (same source as Fig. 4)

References

1   LEACH, G.: 'The motor car and natural resources'. Division of Urban Affairs, OECD, Paris, October 1972

2   OECD Europe covers the countries normally included in Western Europe, plus Turkey, but not Yugoslavia

3   BERRY, R.S., and FELS, M.F.: 'The production and consumption of automobiles'. Report to the Illinois Institute for Environmental Quality, July 1972

4   'Motor vehicle indicators 1955-1985'. OECD, Paris

117

3.4   Energy and future transport

C.J. Clemow
British Railways Board, London, UK

The problems arising in the future for transport that are related to energy considerations have some special features not generally found in other uses of applied energy. The accumulating evidence is that, although total energy resources are relatively large — particularly if one includes solar power — the readily portable fuels, oil and liquefied gases, are being rapidly depleted. However, the latter are the fuels on which transport depends, and they are likely to be in very short supply well before the end of this century.

The changes that will follow will act in two ways. First, the direct solutions to the shortages of the existing cheap and convenient fuels will increase the cost of transport: the solutions can include paying the rising price for the same fuels, synthesising equivalent fuels, and employing less convenient and more expensive fuels. Secondly, the difficulties, solutions and rising costs will fall unevenly on different modes of transport, so that the shares of total transport demands carried by each mode will change, and more traffic will be carried by those methods finding themselves at the least disadvantage in the new situations.

In general, it can be concluded that those forms of transport using fuel most efficiently will tend to succeed better than now, and that, in all forms of transport, there will be new inducements to seek fuel economies and substitution of cheaper fuelling methods.

Several factors determine the relatively good or bad use made of energy in transport. Some of these are combined in Table 1, comparing fuel consumptions by road and by rail. The figures are qualified by the knowledge that the capacities available in each type of vehicle are only partially used. For example, average seat occupation in private cars is around $1 \cdot 2 - 1 \cdot 5$ (about 30% of capacity), whereas railway trains typically have average load factors in excess of 50%. The average speeds are also very different between modes, as also the types of duty, frequency of stops and so on.

Table 1

Comparison of fuel consumptions in road and rail transport

Freight	Capacity : ton miles per gallon of diesel fuel
Road : vans up to 2 tons	$25 - 50$
: trucks up to 8 tons	$50 - 110$
: truck over 8 tons	$100 - 160$
Railway trains	$160 - 320$

Table 1 (Continued)

Passenger			Seat miles per gallon of fuel
Road	:	private cars	100 − 120
	:	buses	380 − 560
	:	coaches	480 − 600
Rail	:	express trains	200 − 280
	:	suburban trains	300 − 420

From the Table, it can be seen that larger transport units are more efficient, as might be expected. Railways should secure an additional benefit from the lower resistance inherent in steel wheels running on railways compared with rubber tyres on roads. It may therefore be surprising that buses and coaches compare well with passenger trains.

Table 2 compares typical weights and capacities of road and rail vehicles, and it can be seen how comparatively poor the railway train looks against the bus and coach counterpart (one version of the Advanced Passenger Train will raise the ratio of seats to gross tons up to 1·7). The reasons for the large weight of trains are threefold. The average speeds of railway trains are higher than their road counterparts; a much higher degree of space, comfort and facilities are provided, including refreshments and toilets; trains are constructed strongly to give considerable safety even in a major accident. For the future, it must be probable that a concern for fuel economy will engender design techniques to reduce the weight of trains, perhaps even compromising the three special features mentioned above.

Table 2

Comparison of weights of typical road and rail passenger transport

Type	Seats	Gross weight, including passengers	Seats per gross ton
		ton	
Private car	4	1·1	3·6
Coach	50	10	5·0
Bus	64	11.4	5·6
Express train	540	490	1·1
Suburban train	740	350	2·1

The use of electricity is a common feature in transport vehicle transmissions, but a special case is that in which power is supplied through an external conductor system, as in electrified railways, trams and trolley buses. The latter seem likely to have a new importance in the future. For a long time after the impending shortage of portable fuels starts to press in (and perhaps indefinitely), there will be continued availability of other fuels such as coal, nuclear materials, and solar energy, which will enable electricity still to be generated reasonably cheaply. Hence the direct use of electricity in transport will become much more attractive. However in their present forms, the flexibility of the conventional road vehicle that can go anywhere is not available, owing to the need for mechanical (usually sliding) contacts between the vehicles and the fixed conductor system.

One means of overcoming this disadvantage is to use batteries, and, no doubt, battery vehicles will be used in many more applications. Present-day efforts to promote such vehicles have not succeeded commercially, because the fuel cost (including replacement batteries) has been very high compared with the internal-combustion engine using petroleum fuels at their present low prices. Apart from special applications like milk-delivery vans and industrial tractors, perfectly workable versions of battery-electric cars, buses and trains have also been built and operated.

Although the theme of this paper is concerned with energy, it should also be remembered that many possible future applications of man's ingenuity in engineering may be as much inhibited by shortage of other resources, particularly metals, as it may be by shortages of particular fuels. In transport, this would indicate the need to use systems that, while providing the desired characteristics in amenity and performance, have low gross weights and low energy consumption preferably derived from general-purpose electric power stations. The broad conclusion is to foresee a trend of change to greater use of public transport by road and rail and increasingly employing electric traction. It is fortunate that this harmonises with the growing popular demand to reduce road congestion, noise and air pollution caused by road vehicles.

3.5 Long-term energy requirements for ships and aircraft

J.E. Allen

Hawker-Siddeley Aviation Ltd., Kingston upon Thames, UK

1. Introduction

Of all the classes of future prediction, that of deciding the balance between the population's needs and the resources available is probably the most difficult (References 1 and 2). In assessing future energy requirements for these two important travel media it is first necessary to assess how world traffic may increase over several decades, and secondly, to assess the likely share these systems will take of the total traffic (Reference 3). In looking so far ahead some new engineering solutions are to be expected which may be more appropriate to a future era very different from our own (Reference 4).

Transport may be subdivided into freight and passenger categories. In the former, there is on the one hand 'pipeline' delivery, such as oil tankers, and, on the other, specialised urgent freight of diverse characteristics. The former may be carried more in the future in actual pipelines, and the latter will increasingly be carried by air. Passenger traffic arises from several human activities, e.g. business, emigration, leisure, cultural, diplomatic, and even criminal. At present there are many transport systems, e.g., road, rail, ship, air, hovercraft, and even spacecraft. Improvements in efficiency, convenience and pollution aspects are expected to continue as old systems wear out or are replaced by more suitable forms, e.g. the people mover in the cities and the quiet V.T.O.L. aircraft between cities. Traffic growth depends critically on economic factors, e.g. fares must be reduced to attract significantly larger numbers of tourists. Transocean passengers have largely abandoned shipping for the airlines, and, in their turn, some of the business passenger traffic over long distances may be replaced by sophisticated telecommunication links.

1.1 Traffic growth

This is notoriously difficult to predict, and, in a period of growth, is usually under-estimated. Often it is estimated by taking present day values, extrapolating along the trend of the last few years and placing maximum limits deduced from other desiderata. One example is the 'gravity model', in which intercity traffic is proportional to the product of the populations of the cities and inversely proportional to some power of their distance apart. In setting limits to energy supplies in the far distant future the expected world population is an important factor. Low estimates suggest 7 billion by 2050 AD (with birth control) and 13 billion (and increasing rapidly) without birth control. Doxiadis (Reference 5) quotes a maximum population of 50 billion by 2120 AD when 49 billion would live in urban areas.

Other estimations on traffic growth have been based on observation over 25 years in the USA, which showed that transport expenditure could be related directly to other consumer expenditure. More recently, however, there has been a continuously increasing demand for travel showing a distinct consumer priority for transport over other forms of spending. This 'need' to travel can be attributed partly to man's natural curiosity and partly to a yearning to escape from the increasing overcrowding and tension, making holidays a necessity. It is thought that in the future this need for travelling will be increased with the increasing strain of city life, and telecommunications will stimulate man's curiosity rather than replace it (Reference 6).

The average American travels 3% more miles and spends 3·8% more on transport every year. This increase in cost per mile is almost certainly due to a desire for increased comfort and speed in travel. Between 1950 and 1969, UK sea passengers increased by 4·0% per annum; air traffic by 14·2%. In 1969 the totals were 9·6 million and 17·7 million passengers, respectively. In general traffic values are increasing more rapidly than the population (Reference 7), see Fig. 1.

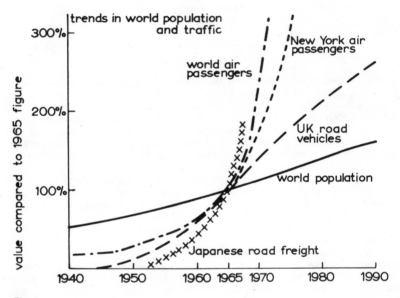

Fig. 1

1.2 Transport share of energy resources

The consumption of energy in the USA (Reference 8) in 1970 was distributed as shown in Table 1.

Table 1

Energy use, USA (1970)

	Percentage
Automobile	18·0
Other transport	6·6
Industrial	37·2
Residential and commercial	22·4
Conversion and transmission losses	15·8
Total	100·0

122

If petroleum (OECD 1970) is considered, a similar table can be derived where transport takes 43% of the total, of which 31% is used on the roads and only 5% each by sea and air. Similarly, in the table above sea and air only represent 2% each of the energy consumption.

The above table could be taken as being representative of a developed nation of the future, assuming no radical changes.

The world energy consumption is $6\cdot6$ x 10^9 KWh per year. In the USA this represents 10KW per person per day; the rest of the world only used $0\cdot1$ KW per person per day. By 2020 AD it is estimated that man-made energy may account for $1\cdot1$ x 10^{11} KWh per year.

Although it is not strictly in the scope of this paper, it is immediately apparent that too much energy is being used by the motor car in particular and by the USA in general. It should be determined if one is directly affected by the other, and the overall economics of the motor car should be examined in detail. A better use of state owned transport may lead to a slowing down of the usage of valuable resources, and this step may be necessary in the advent of a fuel shortage. Also it should be determined whether a high standard of living and high energy usage are necessarily linked.

1.3 Assumptions on energy supplies

There are essentially two groups of energy. Regenerative energy — direct solar energy and its derivatives, such as food, wood, peat, water, wind, etc., and non-regenerative energy — fossil fuels and nuclear energy. The total world resources are predicted (Reference 9) on a knowledge of the resources in a certain area applied to areas of similar geological history (Reference 10). Certain resources, particularly oil and uranium, are difficult to estimate because of company, military and political secrecy.

Energy consumption and sources can be compared using a basic energy unit of 'tons of coal equivalent' (tce), where a basic annual energy unit is taken as 10^{10} tce, see Table 2.

Table 2

Energy consumption/resources (Reference 11)

	tce p.a.
Present world energy consumption	6 x 10^9
Predicted world energy consumption (1982)	1 x 10^{10}
Solar energy received by earth	$1\cdot5$ x 10^{14}
(energy lost by radiation from earth)	
Primary photosynthesis	$1\cdot5$ x 10^{11}
Economically exploitable fossil fuel	$3\cdot4$ x 10^{12}
Hydroelectric power (absolute world total)	8 x 10^9
Cheap uranium in earth's crust (fission reactor)	1 x 10^{10}
Cheap uranium in earth's crust (breeder reactor)	50 x 10^{10}

The world's energy resources can alternatively be regarded in terms of weight, there being $7\cdot64$ x 10^{12} metric tonnes of coal able to produce $5\cdot6$ x 10^{16} thermal KWh, $2\cdot7$ x 10^{11} metric tonnes of petroleum, and between 3 and 20 x 10^{15} tonnes of carbon in limestone, oil shale, and other sediments.

It is estimated (Reference 12) that fossil fuel reserves will be consumed by 2900 AD in Table 3.

Table 3

Fossil fuel consumption (Harrison Brown)

Year	1900	2100	2300	2500	2700	2900
Tons per year (billions)	1	19	19	8·5	2	0

Fuel consumption has been rising continuously, but it was not until 1967 in the USA that this rate overtook the gross national product, which indicates a less efficient production of energy.

Efficiencies of various forms of energy use are compared (Reference 8), see Table 4.

Table 4

Conversion of energy (Summers)

Open fire	20%
Furnace	75%
Fossil fuel space heating	50-55%
Electrical energy conversion	33%
Incandescent lighting	5%
Fluorescent lighting	20%
Automotive engine	25%

It is worrying that two of the most inefficient energy converting processes are the most used. It is estimated that lighting consumes about 24% of all energy produced and the automobile uses about 25% of the energy budget. If cars were all changed to have electrical propulsion the effect would still be about the same as the energy would have merely changed sources to the power station.

By 1980 it is estimated that 25% of the electric power will be produced by nuclear reactors, rising to 50% by 2000 AD. Peak electric power will be produced by gas turbines coupled with generators in this system.

Alternative sources could utilise the energy from tides and winds, (quite feasible study has been put forward using the latter) and storing the irregular energy by using it to decompose water to hydrogen and oxygen and using them when required. 55% efficiency is estimated. France has already built a 240 MW power plant using tidal energy.

A fact which is disturbing is that in an exponential growth, such as that experienced in energy demand, in a period of doubling demand, the increase is equal to the total amount supplied in its history. It is estimated that by the year 2000 AD the energy requirements will be triple those of today, requiring a tremendous advance in energy producing technology.

2. Sea transport

2.1 Sea traffic

The majority of the sea traffic is at present ocean-going oil and raw material transport (Reference 13), passenger and coastal traffic play a much smaller but important part.

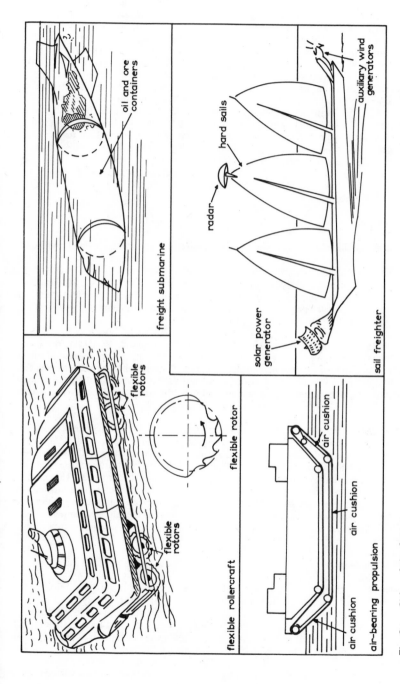

Fig. 2 Ships of the future

freight submarine

oil and ore containers

radar

hard sails

auxiliary wind generators

solar power generator

sail freighter

flexible rotors

flexible rotors

flexible rotor

flexible rollercraft

air cushion

air cushion

air cushion

air cushion

air-bearing propulsion

125

Oil transport, being a valuable cargo, is a major item. In 1970, when oil production growth rate was about 5% per annum, nearly 50% of the production was being taken by sea, representing 1100 million tons, in oil tankers totalling over 175 million tons in deadweight (Reference 10). However, tankers are uneconomically powered, consuming between 5 and 10% of their oil in transit, but their advantage lies in their size and ease of pumping oil in and out. The traffic pattern is sensitive to ship availability and to continental needs which could change significantly with the discovery of new hydro-carbon fuel sources off-shore in Europe, Canada and Australia (Reference 14). The total traffic flow may be influenced by the construction of new waterways, such as the Thailand Peninsula canal, but is not expected to increase by orders of magnitude (Reference 15). Estimates put the growth to between three and four times the present capacity by the end of the century; much of the traffic growth being accounted for by increased air transport. A recent Department of Trade survey in the USA predicted a growth from 2 billion tons in 1970 to 8 billion tons in 2000 AD for world sea trade. This volume of traffic would require 2500 ships of at least 30 MW (40 000 hp) by 1990, and a NASA report shows that by 1980 nuclear powered ships of this size will be as economical as conventional tankers; the Japanese plan to have 280 nuclear ships by 1990 (Reference 16).

2.2 New concepts

A recent searching look into sea operations, taking into account the overall picture of source to consumer economics showed the current ships to be uneconomic (Reference 13) The situation could be vastly improved by a growth in size and improved packing by containerisation and palletisation. It was shown that ports are not the best places to pack, repack and store goods. A more rapid turnround leads to a higher utilisation, and hence more efficient operation. The size growth would only be limited by costs of dredging ports, storage and insurance of the vessel.

Should the need for the rationing of fossil fuels occur around 2200 AD say, replacement of alternative energy sources would be easier in ships than in aircraft because of the less severe space and weight penalties. Already, nuclear powered ships have been operated safely if not economically. Nuclear fusion has not yet been demonstrated; the systems being proposed being more applicable to land use. It is rather premature to make any predictions to maritime use although this seems highly desirable.

Inventions which could change ships characteristics include (Fig.2), Thring's moving belt propulsion (Reference 1), Kearsey's buoyant rotor ships (Reference 17), and nuclear powered submarine freighters. The development of such alternative systems may reduce fuel consumption for a given volume of traffic by up to 30%, but the vehicle speeds are likely to be limited by the Karman-Gabrielli relationship of 1950 (Fig.3). The most likely of the three above inventions is the nuclear submarine which has the advantages of avoiding surface wave friction and uses the increased speed obtained to achieve a greater utilisation and hence a better economy. It also has the added advantage of operating in polar waters, which could be useful as Canada is a major iron ore and oil supplier.

2.3 Comment

With the growth of ships, perhaps one of the most important factors is collision avoidance (Reference 18), which may impose some restrictions. Energy needs in the next century are likely to be restricted if not critical, and a change to nuclear power could reduce the rate of exhaustion of the depleted supplies of oil fuel. In the event of a total lack of energy resources the return of a modernised mammoth sailing ship using advanced technology in the design of aerodynamic hard sails of variable camber, new structural materials and radar, is possible (Fig.3). About 0·2% of the solar energy reaching the earth appears in winds, giving a potentially available energy source in the order of 10^5 MW.

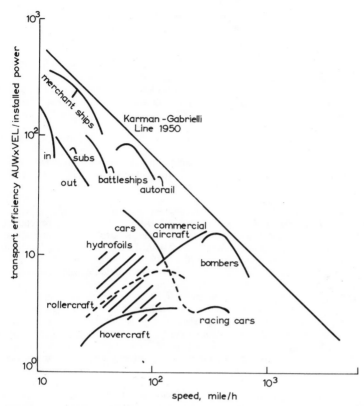

Fig. 3   Transport efficiency and speed

It could therefore be concluded that the problems of sea transport, arising from long-term shortages of fossil fuels, could be contained within practicable solutions.

3.   Air transport

3.1 Air traffic

One source (Reference 3) gives a value of $2 \cdot 3 \times 10^{11}$ passenger kilometres flown in 1967 and an increase in traffic of 14% per annum, giving $1 \cdot 6 \times 10^{12}$ and 14×10^{12} passenger kilometres flown in 1982 and 2000 AD, respectively.

Other sources believe that noise and airport saturation will have a marked effect on the demand, and the growth rate will decrease to 9½%, with noise effects and to 4% when congestion affects the passenger demand.

In the USA for the period 1968 to 1971 there was no growth rate in air transport although the rest of the world's traffic continued to increase. Future demands are, however, forecast to increase rapidly to triple the present level by 1982, and assuming a load factor of 55% this shows the present capacity, existing and on order, to be able to cope with demand only until 1974. This capacity is not, however, evenly distributed (Reference 19), and the obsolescences of old aircraft probably worsens the situation.

Air freight traffic is rising faster than passenger traffic at 16% per annum, representing 0·04% of the world cargo tonnage, raching 4 x 10^9 tonne-kilometres in 1965 and rising to 3.2 x 10^{11} tonne-kilometres by 1985, 5·6 x 10^{12} tonne-kilometres by 2000 AD.

Thring (Reference 1) envisages an increase in passenger miles flown by 25 to 50 times within the next century, implying an aircraft fuel consumption of 10^9 tons per annum. An alternative 'ultimate' criterion in air transportation is that each person flies one inter-continental flight in their lifetime, giving an increase in traffic of over 10 times the present value at 0·5 x 10^{13} passenger-kilometres.

3.2 Possible limitations

Although the total atmospheric volume is capable of holding over 1000 times the present air traffic flow, other restrictions are far more stringent, and this could lead to a fall in growth until a plateau of demand is reached.

Airport congestion is already familiar, with noise and pollution annoyances (Reference 20). Because of the existing designs of aircraft and airports, and the need to separate aircraft by a safety margin, this situation will continue and problems will become serious enough to demand remedies long before 2000 AD. At London's airports a 200% annual passenger increase is expected between 1970 and 1980 with a corresponding 380% growth of freight traffic (Reference 21). Larger aircraft should increase passenger volume within existing traffic clearance, and fully automated flight control could increase the rate of landing and take-off at airports. However, orders of magnitude improvement are unlikely to be obtained with existing types of airports, and increasing numbers of airports will not alleviate the problem entirely because of air traffic complications. New systems must be introduced using computer controlled air traffic systems and alternative ways of providing airports. S.T.O.L. (Reference 22) providing distribution of air traffic to regional centres and vertical take-off and landing into city centres are two ways that will contribute really significant improvements.

Prohibition of night flights because of aircraft noise puts a large restraint on the growth of traffic, and is uneconomic, from the point of view of the operators, as the aircraft is not earning any money for half the day (Reference 24). Quieter aircraft now being built will significantly reduce the noise intensity, but it is unlikely that all of these would be allowed to fly during the night. However, they will not have to throttle back on take-off to reduce the noise level, which has been the source of, or contributed to, several accidents. It would need a radically new method of propulsion to eliminate the noise completely (Reference 25). Some of the novel systems have been rejected in the past because of operational economic factors, but might become feasible if this could be overcome. Again the vertical take-off aircraft would help with this problem of noise, as the noise footprint is much restricted in area.

Before long it had been observed that tourist resorts tend to degenerate into a universal Coca-Cola syndrome. Saturation will only be avoided before 2000 AD if there is an alteration in the fast growing tourist traffic. If leisure time is increased so that rapid transit is not essential, a return to sea cruising is possible.

In the 1960s airlines made reasonable if not outstanding returns on capital investment, but since then profits have dwindled away to the point that many airlines are in substantial debt. This they owe to inflation and sensitivity of the market to political and other fluctuations. Hijacking has also had its effect with delays, extra costs and financial uncertainty. New and more efficient airliners need substantial launching costs, and even 25% improvement in direct operating costs reduces to less than 10% when all systems and traffic factors are taken into account. It is quite impossible to forecast such delicately balanced issues over many decades, but some authorities see the future as allowing only 10 or so airline/aircraft manufacturing companies throughout the world with a financial situation such that an aircraft is made specifically for the airlines which will provide the money for the project.

3.3 New concepts

The present trend towards larger aircraft such as the Airbus and Jumbo jet may be extended, especially in the shorter range commuter market in which studies for enlarging the 370 seat 4700nm Boeing 747 to a 750 seat 2700nm version, are being considered (Reference 26). Aircraft increasing further in size are also envisaged, such as the 3·5 M lb resource carrier and the nuclear powered 11·75 M lb aircraft (Fig.4).

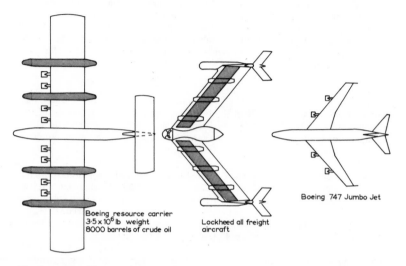

Boeing resource carrier
3·5 x 10⁶ lb weight
8000 barrels of crude oil

Lockheed all freight aircraft

Boeing 747 Jumbo Jet

Fig. 4 Two Mega projects

Aircraft propulsion in the future will utilise the new generation of high by-pass ratio quiet turbofans reducing the noise to an acceptable level, and also reducing the weight and specific fuel consumption of the engine.

Supersonic aircraft are likely to take up 25% of the world's international traffic essentially on the longer oversea routes (Reference 27). Hypersonic transports in excess of Mach 5 have been studied for flights of 8 - 13 000 km (5 - 8000 miles) using liquid hydrogen (Reference 28). However, inherent difficulties especially with noise make it unlikely to be operational.

Manned spacecraft seem to have no advantage over conventional flight for terrestrial transport, being used solely for the space shuttle in orbit, as discussed in the next section.

The present day interest in the revival of the airship (Reference 27) could be accomplished with modern methods of technology (Reference 29), and it is envisaged that a fleet of large, safe, lighter-than-air craft could take over a significant proportion of air transport, with fuel consumption 1/15th of an aircraft carrying the same load over an equal distance (Reference 30). However, the cost of aircraft operations depends on the time of flight, but revenue is proportional to the distance flown, and so, unless fuel shortage forced up fuel prices by an order of magnitude, the airship is likely to remain on the sidelines except for special purposes.

The present family of aircraft can be made more efficient by reducing or enlarging the aircraft size according to demand, by utilising the high by-pass engine, by using super-critical airflow wings, by using the new composite materials and by using stability augmentation to reduce pilot workload.

3.4 Comment

Since most of the component efficiencies in transport aircraft are close to the theoretical maxima, no improvement in aeronautical techniques could significantly reduce the large orders of magnitude increase in fuel consumption implied by the expected traffic growth. No means of supplying energy needs for aircraft other than liquid hydrocarbons are at present discernable. Clearly liquid hydrogen could be used as a fuel, obtained by electrolysis of water in conjunction with nuclear fusion. However, although liquid hydrogen is used for manned spacecraft, and there seems unlikely to be any tremendous obstacle in making it ultimately safe for passenger transport, it would not be generally acceptable because of its low density and the high drag of its tankage. It does not so far appear to be sensible for subsonic or low supersonic speed aircraft. Presumably the effect of water vapour exhaust in the high stratosphere would need to be studied, but is probably more benign than hydrocarbon exhaust.

Nuclear aircraft have been proposed of 2M lb gross weight, being able to utilise half this weight for cargo. In a ten hour flight the fuel consumption would be less than ½ pound. The problems associated with this kind of aircraft are mainly concerned with safety, especially in the event of a crash, and nuclear power is probably far better employed in producing hydrogen, and hence synthesising hydrocarbon fuel.

Thring has proposed beaming coherent magnetic radiation to propel aircraft (Reference 1), but at present there seems no way of transmitting the enormous power required. Energy transference could not be by conventional electrical motors, and some more direct means of converting electrical power to air pressure would be needed. There is also a major problem in organising the complex ground equipment to direct the power by discrete beams: power failure would lead to disastrous results.

With the approach of fossil fuel exhaustion it may be that hydrocarbons generated from botanical substances could be used. It is more likely however that the world food shortage at that time would claim greater demands than transportation.

Another transportation solution more in the science fiction trend has been put forward by A.C. Clarke (Reference 31), following proposals to cover cities by a geodesic dome, who proposed that heating of the air inside the dome would give it enough buoyant lift to allow transportation of the whole city. This may be possible in the future, but it has incredible problems especially in stability and control.

4. Space

Space flight's contribution to traffic, fuel usage and communications can be divided into four main areas:

(i)   scientific satellites for observation of the atmosphere and surface conditions as part of the Earthwatch programme

(ii) communication satellites which should revolutionise education in some countries and could be important in spreading conservationist attitudes

(iii) some reduction of long distance vehicular traffic by permitting complex data exchange, e.g. working conferences by TV and at the maximum utilisation, e.g. enabling medical specialists to perform surgical operations by radio controlled remote manipulators. It is impossible to forecast whether such activities would significantly reduce overall traffic volume.

(iv) solar generators in stationary orbit. A panel five miles square could accept $8 \cdot 5 \times 10^7$ KW, which allowing for 18% efficiency and transmission losses allows an earth station six miles square to receive 10^7 KW of electrical power.

(v) NASA space shuttle. This assumes that there will be 74 operations per year on average, carrying 3000 tons of payload per year from 1978 to 1990, and using 70 000 tons of propellant per annum. This represents only an insignificant drain on fuel supplies.

Long range ballistic rocket passenger transport is not considered to be a viable contender for the future. The flight time would certainly be short − forty minutes for 10 000 miles, but the noise and accelerations imposed on passengers seem inherent in the principle of the system which could no doubt be built if required. To be socially acceptable a quiet system of propulsion would be needed, which is as yet unknown.

If the Earth became uninhabitable, then manned space flight would become a necessity in a mass emigration to another planet, but before then other modes of transport are likely to have higher priority for fuel supplies.

5. Problems

5.1 Commissioning radical changes

It has been said that inventions cannot be bought or managed, and yet the inventors of the past must have taken some account of human needs, e.g. better transport or power, or have envisaged a probable application. It is very rare, however, for any invention to be put immediately into use. Masefield has drawn attention to the interval of between 16 and 17 years needed to bring a new engineering system into accepted use (Reference 32), see Table 5.

Table 5

The developing technology interval (Reference 32)

Innovation	Demonstration		Realisation		Interval (yrs)
Aviation	Wright Bros.	1908	First Commercial Flight	1919	16
Gas lighting	Experimentally	1792	Street lighting	1810	18
Electric light	Incandescent lamp	1878	Theatre lighting	1896	18
Motor car	Benz	1887	Ford production	1905	18
Rocket	Liquid rocket	1926	V-2 against London	1944	18
Jet	First flight	1939	Commercial service	1958	19
High speed flight	600 m.p.h.	1945	Schedule service	1962	17

Different long term objectives in aeronautics may be discerned having a somewhat longer time interval. Some possible 'epochs' are shown in Table 6 as one indication that over long time intervals there can be noticeably different objectives. These objectives are usually hardly noticed, but play a part in co-ordinating research, design and resources towards a goal. To accomplish the 2000 AD target would require considerable readjustment of research and realignment of a nation's funding sources as well as international regulations, and so may not be too pessimistic a date. It also allows for a 'Masefield' technology interval to occur in the period. The 2030 AD target will require an international policy of energy rationing or consumption restraint of a kind never before seen

Table 6

Possible epochs in aeronautics

Period	Objective	Solutions
1910	First demonstrations of possibilities	Sport flying
1940	Minimum weight – maximum performance	Light alloys, new fuels superchargers
1970	Minimum operating costs	Jumbo, fan jets
2000	Balanced overall system/society criteria	Quiet flight, non-polluting exhausts. VTOL
2030	Optimisation within a global energy policy of restraint	Nuclear airships, hydrogen fuelled aircraft with laminar flow and wake propulsion systems.
2060/ 2300?	Global invariant energy systems	To be specified, invented and developed.

on Earth. This aspect is likely to take longer than that needed to engineer the new systems. The last target – the Global Invariant System, poses the greatest problems for engineers. If it could be successfully accomplished it would automatically be accompanied by consumption restraint, and some thought is required to imagine how this would be regulated between nations and interested organisations.

5.2 Conservationist issues

From the above discussions some points immediately come to notice:

(i) Will the population growth continue as it has done over the last 50 years? The Rev. T. Malthus predicted in 1798 that there would be checks to the population growth (Table 7), many of which are applicable today with added restrictions (Table 8). Possibly also the degeneration of cohesive drives and constructive societies could lead to loss of fertility in the future.

Table 7

MALTHUS 1766-1834
Checks to population growth

Moral restraint	Excesses of all kinds
Vice	Diseases
Unwholesome occupations	Epidemics
Severe labour	Wars
Exposure to the seasons	Plagues
Extreme poverty	Famines
Bad nursing of children	Celibacy
Large towns	Infanticide

Vicious practices (i.e. birth control)

Table 8

Population checks 1972

Atmospheric pollution	Death from overfeeding
Genocide	Death from smoking/drugs
Nuclear wastes	Transport accidents
Guerrillas and hijackers	Birth control
Suicides, mental and stress illness	Abortion

(ii) There may be additional demands on energy sources in the future. Water shortage may become critical before the turn of the century, and hence desalination plants will become operational, absorbing significant amounts of energy.

(iii) Mere statistics of volume of traffic do nothing to clarify whether the amount of travel is really worthwhile. Perhaps a great deal of travelling is quite unnecessary and a more leisurely age might emerge which did not feel such an urge to travel so much or so quickly.

(iv) The promise that TV comsats for conferences, and remote operations would eventually curtail travelling, does not seem to have happened yet. One must be on one's guard against some modern developments which are frequently gimmicky demonstrations and never achieve world wide application. It is also felt that the personal contact is an important feature in communication which cannot be replaced. Nevertheless a marked increase of fuel costs may make such a solution economically viable.

(v) We cannot appreciate the effect of radically new concepts and inventions. Newton studied planetary motions and deduced laws of force and momentum for earth-bound engineering. With spaceflight tools, engineers made extra-terrestrial observatories for astronomers. If observations made in space by these sensors give a new understanding of gravity and nuclear behaviour and a novel way of energy exchange is discovered, much of the present day concern over power and transport may be avoided.

(vi) The atmosphere-ocean-space system interact in a complicated way, and there are several natural safety valves and checks to disturbance of the equilibriums (Reference 33). Will our improving knowledge of the atmospheric responses permit us to anticipate unfavourable trends and allow for them? Is it necessary to try to drive the biosphere back to its 1950 level in an attempt to prevent deterioration? Can we ever really know the capacity of the atmosphere to alter and compensate for man-produced inputs? The large Martian sandstorms suggest that planetary atmospheric motions are little understood.

(vii) Although 'microcybernetic' techniques are very successful in controlling quite complex chemical factories or automatic aircraft flight control systems, we should not fail to recognise the singular lack of success in the realm of 'macrocybernetics', i.e. purposeful control of city-sized, nation-sized and global-sized activities.

(viii) Recognising a possible future state may be totally misleading, since neither the means nor the time is available to accomplish the needed alterations.

(ix) Many of the new global concepts, such as energy rationing or the creative society, require very different controls and regulations which violate established ways of economics. Can we have some 'ultra-economics' which could be seen to be viable in relation to trade and the normal laws of supply and demand?

(x) Conservationism is unfortunately suffering from too many sensationalists and me-toos which regrettably causes over reaction from many who should be enlisted in the quest. Progress will come when those who are truly motivated (Reference 34), and who recognise sensible ways of proceeding (Reference 35) combine with a wider range of professionals in engineering, law, government and business, who can advise and help in their own fields and tackle the problems as they arise.

(xi) Since 1850 combustion of fossil fuels has released 200 billion tons of carbon dioxide (Reference 12). Slightly more than a third of this appeared as an increased concentration in the atmosphere, the remainder being absorbed by plant life and the oceans. It is possible that the biomass has increased by 15 billion tons in the last century. One fifth of the total man-made fossil fuel exhaust emission occurred in the last decade, and hence the rate of increase in the future is a matter of concern. By 2000 AD it is estimated that CO_2 will rise to 400 p.p.m. (290 p.p.m. in 1850). By 2100 AD at the expected maximum rate of fossil fuel consumption it might reach 600 p.p.m. The ocean has an important role in absorbing CO_2, there being 60 times as much CO_2 in oceans as in the atmosphere. The increased CO_2 level in the atmosphere could lead to an atmospheric temperature rise as has been known since 1899. Whether the effect of oceans can keep this temperature rise low is not yet known; but it has been established that volcanic dust and agricultural burning in the tropics will contribute to a slight lowering of the atmospheric temperature. Rejection of waste heat from energy conversion and aircraft condensation trails may have an increased importance in the future.

(xii) Present day society may have got into a state of disenchantment with 'progress' and a feeling of incapacity to control events because of worshipping false gods. These might be seen as:

> Science — to know all
>
> Technology — to do all
>
> Fertility cults — a strange phenomenon of the 20th century
>
> Economics — money solves all

It would be disastrous if the present fervour for conservationism turned out to be a mere pagan worshipping of Earth. Only by seeking the highest level of motives can a tendency to fractionate into discordant voices as at the Tower of Babel be avoided.

6. Conclusions

Table 9

Energy and humanity

> Annual ecological status report
> Society for global responsibility
> Timed achievements program
> Targets for — international regulation
>   — inventors and engineers
>   — government long range policies
>   — public education
> Re-establishment of principles of individual freedom

A Society for Global Responsibility (Table 9) should be established under whose guidance all the discussions of this paper should come. This Society should publish an Annual Ecological Status Report so that the dangers from year to year can be followed and the necessary steps taken. After discussion, a timed achievement programme should be drawn up and effort put into the parts of the programme which are falling behind. This programme should include targets for international regulations, inventors and engineers, government long range policies, and public education. The Society should try to re-state the principles of individual freedom within a civilisation with far more curbs of a general nature in force. It should seek the financial backing needed to promote research into alternative sources of energy and try to restrict the uneconomical use of power.

The targets set out for the establishment of such a Society are difficult, to say the least, and imply a powerful organisation. Such is the nature of man that it is doubtful whether government will respond swiftly on such issues, except in an emergency, but unless they do they will be faced with such an emergency. It is therefore essential that such a society as has been proposed is given the necessary power and encouragement.

7. References

1. THRING, M.W.: 'Power generation for aircraft in the second-century', The Aeronautical Journal of the R.Ae.S., September 1968

2. MALTHUS, T.R.: 'Essay on the principle of population', 1798. BONAR, J.: 'Malthus and his work', London, 1924. KEYNES, J.M.: 'Robert Malthus, Essay in biography', 1933

3. BOULADON, G.: 'The total transport demand' in 'Aviation's place in World transport', R.Ae.S., May 1971

4. CLARKE, A.C.: 'Profiles of the future', (Pan Books, London, 1962)

5. DOXIADIS, C.A.: 'Ekistics' (Hutchinsons, London, 1968) p.215

6. CARTRIGHT, E.M.: 'The new civil aviation within our grasp', Astronautics and Aeronautics, January 1972, p.30

7. DANFORTH, P.M.: 'Transport control' (Aldus Books, London, 1970)

8. SUMMERS, C.M.: 'The conversion of energy', Scientific American, September 1971, p.149

9. HUBBERT, M.K.: 'The energy resources of the Earth', Scientific American, September 1971, p.61

10. LUTEN, D.B.: 'The economic geography of energy', in 'Production and shipping of oil etc.', Scientific American, September 1971, p.165

11. ROSENQUIST, I.T.: 'The World energy resources', Society for Social Responsibility in Science Conference on 'Energy and Humanity', September 1972

12. BOLIN, B.: 'The carbon cycle', Scientific American, September 1970, p.125

13. HURST, R.: 'The work of the British ship research association' (Mammoth tankers, bulk carriers and fast container ships), Proc. Inst. Mechanical Engineering 1972, **186**, 5/72

14. 'BP statistical review of the World oil industry', The British Petroleum Co. Ltd., London, 1971

15. LITTLE, A.D.: 'Large-scale intercontinental shipment of fuels', Bulletin, No.495, July-August 1972

16. ROM, E.E.: 'What can nuclear energy do for society?', Astronautics and Aeronautics, January 1972, p.56

17. KEARSEY, J.A.: 'The Rollercraft', Brighton Polytechnique thesis

18. 'Collisions at Sea: Tanker manoeuvrability', The Times, Saturday, October 31st 1970

19. 'Long term regional trends total tonne-kilometres performed by scheduled services of airlines registered in Icao states of each region 1952-1971'. Annual Report of the Council, 1971, Doc.8982 A19 - P/1, p.10

20. GUNSTON, W.T.: 'Crisis in air transport', Science Journal, January 1968

21. WHEATCROFT, S.F.: 'Air Traffic in the 1980's,' Journal of Institute of Navigation, 1971 **24**, no2

22. MARSHALL, E.E.: 'STOL aircraft in future transport systems', The Aeronautical Journal of the R.Ae.S, October 1971

23. BRENNAN, M.J.: 'Design considerations of intercity V/STOL aircraft' BALPA Technical Symposium, 1970

24. STRATFORD, A.H.: 'Airports and air transport growth and transformation', The Aeronautical Journal of the R.Ae.S., May 1969

25. ALLEN, J.E.: 'Looking ahead in aeronautics', in 'The future of aeronautics' (Hutchinson, London, November 1970) chap.1

26. STEINER, J.E.: 'Aircraft development and World Aviation growth', The Aeronautical Journal, **74**, June 1970

27. LUNDBERG, Bo K.O.: 'Aviation safety, supersonic transports and their effects on society', The Aeronautical Research Institute of Sweden, May 1964

28. KUCHEMANN, D.: 'Possible types of flying vehicle in the future', RAE Technical memo AERO929, April 1966

29. SMITH, R.K. 'C.P. Burgess and the ultimate airship', AAHS Journal, Spring 1969

30. HAFNER, R.: 'Aviation in the ecological climate at the close of the century', Aeronautical Research Council, ARC33 648, April 1972

31. 'Space engineers tackle the problems of cities', Science Horizons, USIS London, 1972

32. MASEFIELD, P.: 'Aviation and the environment — The problem of balance', The Aeronautical Journal of the R.Ae.S., October 1971

33. STEWART, R.W.: 'The atmosphere and the Ocean', Scientific American, September 1969, p.76

34 EHRLICH, P.R.: 'How to be a survivor — A plan to save spaceship Earth' (Ballantine Books Ltd., London, 1971)

35. WARD, B., and DUBOS, R.: 'Only one Earth' (Penguin Books, London, 1972)

OTHER REFERENCES

ALLEN, J.E.: 'The future of aeronautics – Dreams and realities', Royal Aeronautical Journal, September 1971

Large nuclear-powered subsonic aircraft for transoceanic commerce', NASA Technical Memo TMX-2386

STARR, C.: 'Energy and power', Scientific American, September 1971, p.37

OORT, A.H.: 'The energy cycle of the Earth', Scientific American, September 1970, p.54

JANTSCH, E.: 'Technological planning and social futures', Cassel/Associated Business Programmes, London, 1972

ELLE, B.J.: 'Issues and prospects in interurban air transport', Akiebolaget Trycksaker, Norrkoping, 1968

NAYSMITH, A.: 'Population distribution and air transport', RAE Technical Report 69227, October 1969

'Forecasting the future', Science Journal, October 1967

'World design science decade 1965-1975', World trends exhibit, World resources inventory, Southern Illinois University, 1965

ESTES, E.M.: 'Alternative power plants for automative purposes', The Institute of Mechanical Engineers, March 1972

ARMYTAGE, W.H.G.: 'Technological forecasting', The Institute of Mechanical Engineers, 1969

LIVINGSTONE, F.C.: 'Fuels for the seventies – and beyond', Heating and Ventilating Engineer, January 1971

MILLER, R.H.: 'Some air transportation concepts for the future', The Aeronautical Journal of the R.Ae.S., July 1971

DARWIN, C.G.: 'The next million years' (Rupert Hart Davis, London, 1952)

SINGER, S.F.: 'Human energy production as a process in the biosphere', Scientific American, September 1970, p.175

COOK, E.: 'The flow of energy in an industrial society', Scientific American, September 1971, p.135

KATZ, M.: 'Decision-making in the production of power', Scientific American, September 1971, p.191

WARMAN, H.R.: 'Future problems in petroleum exploration', BP Co. Ltd., Petroleum Review, 25, March 1971

Part 4
Possible new developments

4.1 The case for solar energy

P.E. Glaser
Vice-President, Arthur D. Little Inc., Cambridge, Mass., USA

1 Introduction

A riddle popular with French children concerns a farmer, a pond and a water lily. It goes like this. The lily is doubling in size every day. In 30 days, it will cover the entire pond, killing all the creatures living in it. The farmer does not want that to happen, but, being busy with other chores, he decides to postpone cutting back the plant until it covers half the pond. The question is: On what day will the lily cover half the pond? The answer is: On the 29th day – leaving the farmer just one day to save his pond.

The realisation that society will have to find the answer to this riddle has recently surfaced, and the issues of limits to growth are being debated. The complexity of these issues can be illustrated by one of society's most basic needs – energy. The 'energy crisis' generated by the massive production and use of energy in the technologically advanced countries is one of the major topics before the public. As various solutions are offered to overcome the crisis, inadequate consideration is often given to technical development, economic constraints, resource conservation, public health, international trade and politics, consumer protection and social equity. As attempts are made to grapple with the interacting, conflicting and cumulative effects of actions already taken or planned for the future, a consensus is emerging that new research initiatives, institutional mechanisms and criteria need to be developed, combined and co-ordinated to evolve, at first, a rationale for national energy policies, and, in the future, policies to benefit world society.

2 An assessment of energy demand

The use of energy has been the key to the social development of man and an essential component to improving the quality of life beyond the basic activities necessary for survival. The manipulation of energy for the benefit of man has always depended on two factors: available resources and the technological skill to convert these resources into useful heat and work. The patterns of energy consumption can be traced for at least 100 years spanning the industrial era. The salient feature of any chart indicating the history of exploitation of energy sources and their consumption is the almost constant upward march of energy consumption, and, since the 19th century, the gradual advance of energy consumption in relation to population growth.

New fuels have always enjoyed a sudden growth in popularity. But, rather than displacing previously available energy resources, they have merely been added to the conventional energy sources to meet the increases in total energy consumption. Integration of new energy sources into the economy has occurred in reasonably similar patterns for coal in the 1850's oil in the 1900's, and natural gas in the 1930's. Nuclear fuels, which are just beginning to become important as energy resources, also appear to fit this pattern, a pattern which has been remarkably representative of the exponential growth of energy consumption.

The most striking fact about the exponential growth of energy consumption over the last century is that it cannot continue forever. In a world with limited natural resources and a finite ceiling on undesirable interactions of energy-production systems with the environment, the future of energy supply poses a multitude of problems. These facts can be dramatised by considering that, during the next 30 years, the USA will consume more energy than it has in its entire history. Over this time span, the annual US demand for energy in all forms is expected to double, and the annual worldwide demand will probably triple. Projected increases in energy demand indicate that the pressures on energy resources and the environment will be experienced worldwide because each nation will aspire to obtain a larger share of finite resources to maintain and improve the quality of life for its people.

The projected increases in energy demand will tax the ability to discover, extract, and refine fuels in the huge volumes necessary, to ship them safely, to build new electric power stations, and to dispose of effluents and waste products with minimum harm to man and his environment. When one considers how difficult it is at present to extract coal from the Earth without jeopardising the lives of miners, to ship oil without the potential threat of major spillage, to find acceptable sites for new power plants, to control the effluents of fuel-burning machinery, and to dispose safely of radioactive wastes, the energy projections generally accepted to be valid over the next decades indicate the need for a thorough assessment of available options and careful planning of future courses of action.

The need for assessment and planning becomes obvious when one examines the longer-term projections of total energy inputs and compares these projections with the energy inputs to the electric utilities. These projections indicate that, by the year 2020, unless other energy sources are developed, the energy inputs to electric utilities alone will consume most of the natural energy sources available to the USA. These projections underline the need for new approaches to provide alternatives for meeting foreseeable energy consumption.

Major capital expenditures will be required to provide the energy-production machinery to meet the needs of the electric utilities as they plan to provide the electrical power to meet demands. In the USA, these capital requirements are projected to be at least $600 billion over the next 30 years with an additional $30 billion required during that time to carry out the necessary research and development on energy conversion, transmission, distribution, and environmental control (Reference 1). Therefore energy production will continue to require a significant fraction of available capital, indicating that one important limit to the exponential growth of energy consumption will be the rate of capital formation.

The complexities of assessment and planning for future energy consumption can be illustrated by the presently experienced confrontation between the electric utility industry and those opposing the construction of new power plants. The issue that has to be faced in the near future is further development of energy-production methods utilising existing energy resources versus preserving the natural environment in the face of increasing pollution, which, in some areas, is already approaching crisis proportions. The public is demanding substantially more electrical power, and is expecting the power to be available — without shortages or rationing. At the same time, the public is expressing an unprecedented concern about environmental quality, but has not yet faced up to the price that may have to be paid to achieve this quality.

The environmental concerns extend not only to the more obvious factors of products of combustion, waste heat from power plant cooling systems, and possible hazards from nuclear power plants, but also include objections to the use of choice land as sites for power plants and transmission lines from the viewpoints of both aesthetics and economics.

In any assessment of energy supply and demand, the implications of continued economic growth in a closed planetary system will have to be faced. As Forrester states: 'It is not a question of whether growth will cease, but rather whether the coming transition to equilibrium will occur traumatically or with some measure of human intervention which may head off some of the most tragic outcomes' (Reference 2).

New concepts of economic equilibrium within such a closed planetary system are being considered. One concept that is being developed by Mishan (Reference 3) may prove increasingly relevant — that is, amenity rights should share equal status in law and custom with property rights with which they often conflict. Another concept has been advanced by Odum (Reference 4), who suggests that money is no longer an adequate metaphor to describe accurately our various resource allocations and human transactions. The money metaphor needs to be augmented by a system of energy accounting and simulation which could embrace descriptions of how underlying energy-matter exchanges operate and how hidden energy subsidies or outflows obscure or prevent accurate accounting of the real costs, benefits and tradeoffs in human activities.

This is the context in which the various alternative energy resources have to be viewed. Technical and economic feasibility and environmental and social desirability have to be established prior to any major commitment to develop one or more alternative energy production methods, irrespective of the energy source. It is obvious that the opposition to present courses of energy resource development partly stems from the lack of serious consideration of alternatives that may be more compatible with the realities of the closed planetary system of the Earth.

Among the different sources of energy, whether they be nonrenewable, such as fossil or nuclear fuels, or continuous, such as tidal or geothermal, none has a greater potential than solar energy, on which life in its very essence depends. The question that needs to be answered is: Has society reached a level of sophistication to apply solar energy for its overall long-term benefit consistent with the balance of nature? The answer is not yet obvious. However, efforts required to provide the answer are beginning to be made as reviewed in the following sections.

3   The nature of solar energy

The total influx from solar, geothermal and tidal energy into the Earth's surface environment is estimated to be $173\,000 \times 10^{12}$ W Solar radiation accounts for 99.98% of it. The Sun's contribution to the energy budget of the Earth is 5000 times the energy input of other sources combined.

About 30% of the incident solar energy is directly reflected and scattered back into space as short-wavelength radiation. Another 40% is absorbed by the atmosphere, the land surface, and the ocean, and converted directly into heat at the ambient surface temperature. Another 23% is consumed in the evaporation, convection, precipitation, and recirculation of water in the hydrologic cycle. A small fraction, about 370×10^{12} W, drives the atmospheric and oceanic convection and circulation in the ocean waves, and is eventually dissipated into heat by friction. An even smaller fraction, about 40×10^{12} W is captured by the chlorophyll of plant leaves where it becomes the essential energy supply of the photosynthetic process and eventually of the plant and animal kingdom.

Solar energy can be described almost completely by two numbers measuring quality and quantity. The quality of sunlight is such that its thermodynamic potential of theoretical maximum convertibility into work is extremely high. This means that a high fraction of the energy from the Sun is in the form of shortwave radiation capable of photosynthesis, photovoltaic interaction with electrons in solar cells, or coming to equilibrium with high-temperature receivers. The other number is the solar constant of about 1·4 kW/m² outside the Earth's atmosphere.

By the time solar energy reaches the Earth's surface, it is reduced by the various atmospheric scattering and absorption losses to about 1·0 kW/m² on a sunny day. Part of this radiation is received as scattered light on a surface even when clouds are not present. In cloudy weather, the available radiation is greatly reduced, and most of the light received on a surface is scattered light. Solar energy received by a collector includes both direct and scattered radiation, but the scattered radiation does not contribute to the solar energy usefully collected by most energy conversion devices.

The solar energy is further reduced by striking a receiving surface obliquely in early morning and late afternoon and obliquely at latitudes away from the Earth's equator. At latitudes farther from the equator, the shorter day and greater slant of incoming solar radiation reduce the solar energy available in winter from one-half to one-third of the solar energy available in the summer.

The effectiveness with which solar energy can be converted or absorbed on a surface depends on the position of the absorbing surface. The energy received by a surface will be maximum at the equator at noon, because the radiation passes straight through the atmosphere and on to the receiving surface, the minimum length of passage through the air. At all latitudes, the Sun apparently moves from East to West, sweeping through an arc of 15 degrees every hour. In the early morning and late afternoon, the rays pass obliquely through a longer path in the atmosphere, which results in more absorption and scattering. A substantial increase in solar-energy utilisation can be obtained by having the surface follow the Sun in its passage from East to West.

Data are available on the percentage of possible solar energy, the number of daily hours of sunshine in winter and summer, and the number of clear days in many locations. For any location where a major solar-energy application is planned, records of the available solar energy can be obtained to guide detailed feasibility studies.

4   Terrestrial solar energy-conversion methods

4.1   Heat from solar energy

Radiant energy from the Sun is readily convertible to heat; one need only provide a surface on which the solar energy can be absorbed. If the surface is black, more than 95% of the radiant energy is absorbed and converted to heat. If a fluid, such as air or water, is then brought in contact with the heated surface, the energy can be transferred into the fluid and subsequently utilised for practical purposes.

Since the radiation-absorbing surface usually becomes considerably warmer than the surroundings, there is a loss of heat. The magnitude of this loss depends, among other factors, on the temperature difference between the surface and the surroundings. This heat loss can be reduced by (a) removing the absorbed energy at a rate sufficiently high to maintain the surface at reasonably low temperatures (but this may decrease the usefulness of the absorbed energy), (b) reducing the area of surface, while still maintaining the solar-energy input (by use of lenses or focusing mirrors), or (c) providing 'insulation' above the absorbing surface in the form of airspaced transparent materials, such as glass, which permits the passage of solar radiation but greatly reduces the transfer of heat back to the surroundings. Most of the terrestrial uses of solar energy involve one of the two latter techniques (depending on the temperature required) to attain adequate heat-recovery efficiencies.

Theoretically, the heated fluid produced in these solar collectors can be used in any applications that employ conventional fuels. Glass-covered flat-plate solar collectors can deliver heated air or water at typical temperatures of 100° to 200°F, useful in such applications as house heating, domestic water heating, crop drying, and the like.

Focusing solar collectors can deliver comparatively large amounts of energy to small receivers; the receivers can be operated at high temperatures for such uses as steam generation, melting of metals, and heat treating of materials.

4.1.1 Residential and commercial uses of solar energy at moderate temperatures

(a) Domestic hot-water heating

In more than a dozen countries throughout the world, domestic hot water supplies are heated by solar energy. Several million solar water heaters are in use, mainly in Japan, Australia, Israel, the USA and Russia. The system usually comprises a blackened sheet of metal in a shallow glass-covered box occupying 10 to 50 ft^2 of roof area. Water circulates through tubing fastened to the surface of the metal sheet, the warmed water being stored in an insulated tank generally at a level above the solar collector. In a sunny climate, the hot-water supply of an average family can be provided by such a system. Auxiliary heat from an electric heating element or other heat source can be employed if desired. A simpler design involves a transparent plastic envelope, similar to an air mattress, with a black bottom surface. The unit is filled with cold water in the morning; the solar energy absorbed during the day then provides a supply of warm water late in the afternoon. This type has been used extensively in Japan.

(b) Residential heating

The principle employed in solar water heaters can be applied directly as well to the heating of houses with solar energy. By enlarging the solar heating panels — or using more of them — on the roof of a dwelling, sufficient heat can be absorbed in the circulating water to provide most of the heating requirements of houses located in sunny climates. A heat-storage tank must also be supplied, practical considerations dictating sufficient capacity to carry most of the heating demand for one or two typical winter days. Conventional energy sources would be used to supply the balance of the hot-water demand during unfavourable weather.

Air may also be used as the heat-transfer medium in house-heating systems that employ solar energy. Air is circulated across or behind the blackened surfaces in glass-covered panels mounted on the roof, and hot air is delivered by fan and duct work to the rooms of the building. Heat can be stored in water (by use of conventional gas-liquid heat exchangers) or in bins of 1 - 2 in stones (which serve both as heat exchanger and storage medium). When heat is needed in the rooms, air or water can be circulated through the heat exchanger or the bin of hot gravel, the heat then being transferred to the air and to the rooms of the dwelling.

In the past 25 years, more than 20 houses and laboratory buildings have been heated, at least partially, with solar energy on an experimental basis. The USA, Australia and Japan have been most active in this application. Most of the installations have been technically successful, and extensive performance data have been obtained on various modifications of the air- and water-heating systems. Energy savings achieved through heating homes with solar energy can have a significant impact on energy consumption, because, for example, the energy requirements of residential housing in the USA are nearly 30% of total energy consumption.

In most US locations, residential heating with solar energy would be somewhat more costly than present conventional means because of the relatively high cost of equipment still under development. If produced on a large scale, however, solar house-heating systems could be competitive, particularly when all the hidden environmental costs are accounted for. An economic study of residential solar house-heating indicates that, in a least-cost system, about one-half of the total heat supply should be provided by solar energy (Reference 5).

If the solar heating system were designed to supply a large percentage of the total heat requirement, average costs would be quite high, because the cost of solar heating is almost all in fixed capital investment. Because 100% of solar heating is not a rational objective, a combination of solar and conventional heat has to be employed. The optimum combination of solar and electric heat, for example, will be that which is equivalent in cost to the cost of electrical power over a 20-year period.

(c)   Residential cooling

Residential cooling systems that rely on absorption-refrigeration cycles will require more technical development before they can use solar-heated water or air from a roof-mounted collector. However, the solar collector would be the same unit used to heat residences, and the cooling unit would be a somewhat more expensive version of the conventional heat-operated air conditioner. The inherent advantage of solar cooling is that the maximum requirement coincides roughly with the time when the maximum amount of energy is available to operate the system. In addition, the solar collector, which is the most expensive portion of the system, can be employed nearly year-round if cooling is combined with solar heating.

The development of the devices required to cool and heat houses with solar energy has reached the stage where these devices could be available in less than ten years if a mass market is realised. In mass use, solar heating and cooling devices could reduce consumption of conventionally generated electricity by 10%. A step beyond solar cooling and heating would be to replace the black heat-absorbing surface in a solar collector with solar cells. The solar cells would be able to generate electricity in addition to their functions as a heat-exchanger surface.

The technical development of several of these devices has reached the stage where commercialisation will be possible if markets can be identified for each device. Whether the devices are adequate involves consideration of economics, social conditions and cultural preferences. Until now, the evaluation of whether a solar-energy device could meet a need more effectively than a competing device was based on such factors as size, convenience, durability, maintenance, quietness, dependability, and effectiveness in use. Today, however, two important new considerations have entered as evaluation criteria — the absence of environmental pollution and resource exhaustion. The latter two criteria are leading to a re-examination of the potential of these devices and their applications on a scale large enough to benefit from cost reductions inherent in mass production. In the USA this re-examination has led to the view that potentially 10% of new homes could be built by 1985, 50% by 2000, and 85% by 2020 using combined solar heating and cooling.

4.2 Photosynthesis

The maintenance of life on Earth depends on the large-scale application of the photo-chemical reactions of sunlight. These reactions do not require high temperatures to proceed, and theoretically they can be surprisingly efficient. Theoretically, with photo-chemical reactions (yet to be discovered), it should be possible to produce about three tons of product per acre per day. However, nothing remotely of this value has ever been reported. Ordinarily, agriculture produces about this amount in one year. The practical significance of producing agricultural products at this rate is obvious. Should this become successful, then organic matter, including algae, trees and agricultural waste as a fuel souce would have to be reconsidered.

The potential of an energy plantation to grow organic materials for use as fuels to generate electricity in steam-electric power plants has been investigated (Reference 6). Low capital and operating costs compared with other solar-energy-conversion methods, an inherent energy-storage capacity, and environmental compatibility, can be cited as advantages. In addition, nonrenewable energy resources are not depleted, and detrimental environmental effects associated with the use of fossil fuels are largely avoided, because wood contains virtually no sulphur to cause pollution and the wood ashes can be used as fertiliser. The harvesting of the products of the energy plantation will result in smaller environmental impacts compared with strip farming or oil well drilling, and nonconflicting uses of forest areas either for ecological purposes or recreational purposes would be possible.

The land area required for an energy plantation to fuel a 40% efficient 1000 MW steam-electric plant has been projected to range from 300-1100 km^2 depending on solar insulation available at the site. Estimates based on the cost of land, planting and harvesting indicate that fuel costs from an energy plantation would reach $0·45/ million Btu if land costs are $250/acre and harvesting costs are $800/acre. These projections indicate that, where specific conditions are favourable, the energy plantation could be an economical means of harvesting solar energy.

Although photosynthesis in the laboratory can be as high as 30% efficient, it is inefficient in agriculture because of the low carbon-dioxide concentration in air and because of the nongrowing season of winter. In the mass culture of algae, it is possible to obtain photosynthetic yields that are about 20 times those of ordinary agriculture. The algae are grown in large plastic containers with a high concentration of carbon dioxide and artificial cooling.

In addition to food production, enzyme and microbiological processes could be used to produce combustible gases such as hydrogen and methane. For example, if isolated stabilised chloroplasts are coupled to the enzyme hydrogenase, bioconversion of solar energy with the production of hydrogen could be achieved. With a 10% efficient photosynthetic process, 10^7 kWh/month/km^2 could be produced. (A town with a population of 45000 people uses about 32 x 10^6 kWh/month).

Micro-organisms, such as algae or photosynthetic bacteria, when grown on domestic, municipal, industrial or agricultural wastes, in large ponds or fermenters after harvesting would be decomposed with release of nutrients and gases such as hydrogen and methane. Then gases would be collected for use as fuel, and the nutrients are recycled.

The potential exists to combine these processes with integrated utility systems to approach the ideal of a closed ecological system for human habitations in which the demands on natural resources, the production of wastes and undesirable environmental effects are reduced.

4.3 Power from solar energy

The primary advantage of converting solar energy to power is the inherent absence of virtually all the undesirable environmental conditions ascribed to present and expected means of power generation with fossil or nuclear fuels. Several imaginative concepts have been proposed for the large-scale utilisation of solar energy on Earth (References 7 - 10). Which of the approaches will be the most feasible alternative to present power generation methods remains to be established. The important fact is that most of them are based on existing technology and well known physical principles.

All large-scale terrestrial applications of solar energy will require a large land area because of the diffuse nature of solar radiation. This favours the development of energy-conversion devices that can either convert solar energy directly into electricity or, through efficient processes, produce high temperatures.

Essential to the success of any large-scale use of solar energy on Earth will be the capability of storing energy when the sun is not shining, the transmission of electric power over long distances, and the integration of the transmission line into a power-grid system.

The following are examples of concepts that have been advanced and that have the potential to produce power on a significant scale.

4.3.1 Direct energy conversion

This approach is based on photovoltaic conversion of solar energy into direct current by the use of solar cells. This method relies on the direct conversion of photon energy to electricity in a semiconductor crystal, such as silicon. The state of the art of silicon solar cells has progressed, so that efficiencies of 10% are easily reached. In the laboratory, efficiencies as high as 14% have already been demonstrated, and investigations are under way to reach efficiencies of 17%. The theoretical efficiency achievable with silicon solar cells is 22%.

Cadmium-sulphide cells present an interesting alternative to silicon solar cells, because they can be produced by vacuum-depositing materials on a plastic substrate, thus lending themselves more easily to mass-production processes. However, their lack of stability and low efficiencies indicate the need for further development.

New designs include solar cells produced from very-small-diameter silicon spheres, which have been produced from liquid droplets and solidified as single crystals, multilayer cells consisting of several materials, and vertically-illuminated multi junction cells.

Among the materials that are being considered as alternatives to silicon is a gallium-arsenide cell in combination with gallium aluminum alloys, which recently was reported to have achieved 18% efficiency (Reference 11). In principle, there are a large number of compounds that exhibit the photovoltaic effect, and theoretically higher efficiencies should be achievable.

An interesting alternative for direct-energy conversion are the organic compounds that exhibit characteristic semiconductor properties. Several such compounds have been identified. Research on such compounds is continuing, primarily because there does not appear to be a theoretical limit for energy-conversion efficiency. The present efficiency of such compounds is about one-twentieth of 1%, indicating that substantial progress still remains to be accomplished.

The major challenge is to develop mass-production techniques, with consequent cost savings in production of solar cells. Projections indicate that at least a 100-fold cost reduction may be feasible (Reference 12). An alternative way of reducing costs is by partially concentrating solar radiation by means of mirror reflectors to minimise the number of solar cells.

Terrestrial applications of solar cells can be considered for regions where sunshine is plentiful. For example, a 10% solar-energy-conversion efficiency would produce 70 000 kW/km^2 while the Sun is shining. Another suggested application is the use of solar cells on rooftops to provide power to individual residences or, if not required for such purposes, to feed the excess power into an electrical distribution network. Implicit in any of these applications is the need to have energy-storage devices associated with such installations. Typical storage devices are electric storage batteries, pumped-water storage, or the electrolysis of water to produce hydrogen for use as a fuel in internal-combustion engines or in electrochemical fuel cells.

4.3.2 Focusing Collectors

The desirability of producing high temperatures on a surface exposed to solar radiation has long been recognised. High temperatures are needed for efficient operation of engines based on thermodynamic cycles, which can then be used to generate power by conventional means. Focusing collectors based either on mirror reflectors or on Fresnel-lens concentrators can be used to obtain high temperatures. Their drawbacks are that they have to be rather accurately positioned with respect to the Sun and to follow the Sun during the day. The inclination to the horizontal when collectors are fixed depends on the latitude. Focusing collectors have been used in conjunction with small turbine driven power plants. Typically, efficiencies are less than 20%, because focused solar radiation has to be absorbed on a surface through which a fluid or gas is passed that then can be used in a conventional heat engine.

More recently, large solar-thermal plants operating as central power stations have been investigated. Large areas will be required, typically about 30 km^2 for a 1000 MW power station. This requires engineering solutions to transfer the heat obtained at the focusing collectors to the central power plant. Costs and reliability of the apparatus required for this purpose remain to be established. A development that may reduce the complexities of heat collection is based on the use of a self-contained decentralised system for collecting and storing heat obtained from the focusing collectors (Reference 13). Heat pipes would be employed to transfer heat to a small heat-storage tank without the need for a centralised heat-storage facility. Underground pipes would bring water to each storage tank and return the steam directly to turbines.

The long history associated with the development of focusing collectors and data obtained in several installations indicate that solar thermal-power plants represent an interesting option for power generation. The cost of solar thermal-power plants is estimated to be about two or three times the cost of competing fossil- or nuclear-fuel generating plants.

Several promising designs for focusing collectors have been explored; the design inform-ation and performance data available allow a wide variety of design options. One of the drawbacks of focusing collectors is the need to produce optical surfaces to reasonably close tolerances, to maintain highly reflective surfaces for an extended period and to have structural characteristics adequate to withstand wind loads.

4.3.3 Selective radiation absorbers

Another approach to achieving high temperatures is the use of selective radiation-absorbing surfaces, which can overcome one of the major drawbacks associated with heating a surface; i.e. the surface temperature rises as it absorbs the solar radiation until the heat losses offset the heat gain. Usually, the largest heat losses are through radiation of the heated surface in the far infrared; e.g. the maximum radiation loss occurs beyond 5 microns if the surface is heated to 300°C. If the surface could be coated to absorb most of the solar energy in the visible region of the spectrum and emit only a small fraction of the infrared radiation, it would be possible to increase the efficiency of heating with solar radiation by decreasing the heat loss and thereby to achieve higher temperatures.

The desirability of reducing such radiation losses was recognised, and considerable efforts were expended to developing selective radiation surfaces that absorb most of the solar radiation but emit very little thermal radiation (Reference 14). Many different selective radiation-absorbing surfaces were prepared using a variety of tech-niques to coat the surface with very thin layers of materials. At present, the various thin films developed can withstand high temperatures for only limited periods, and exhibit absorption-to-emittance ratios of about 10. In combination with optical concentrators to provide concentration factors of 2 to 4, improved performance could be obtained.

A system based on selective radiation-absorbing surfaces maintained inside an evacuated cylinder, with the heat transferred to a circulating gas or fluid, has been proposed to generate power on a very large scale (Reference 15). Overall efficiencies of about 30% for such a power plant have been projected.

4.3.4 Ocean thermal gradient

Terrestrial solar-energy conversion processes based on thermodynamic or photovoltaic energy-conversion systems have the disadvantage that the large capital investment for the solar-energy converting or collecting surfaces and the associated heat engine-electrical generator power plant can only be used for a portion of each day. In addition, weather conditions further reduce availability of the power plant. One approach that is independent of the daily variation in solar energy is based on the thermal storage represented by Sun-heated oceans.

The concept of using the temperature difference between Sun-heated upper surface layers and deep cold water in the tropical oceans to produce electrical power and fresh water was first suggested in 1882. An experimental plant was built off the coast of Africa, and, in the 1920s, in the Carribean. The latter plant was destroyed by a hurricane before it became operational. More recently, improved techniques have been suggested for such a plant (Reference 16).

In operating such a plant, warm surface water is circulated through a heat exchanger, where it heats and boils propane at high pressure and a temperature of about 22°C. The propane vapour expands through a turbine driving a generator. The propane is exhausted into a condenser and liquefied at 11°C. The heat of condensation is transferred to cold water pumped from a depth of about 600m. The liquid propane is then pumped back to the boiler. The plant is located on board a ship anchored in the most desirable location. Designs for such a plant project power outputs of 500 MW. Electrical power would be transmitted to shore by under sea cables.

The gulf stream represents a nearly inexhaustible reservoir of heat, and a large number of installations could supply a significant power demand or via electrolysis of sea water produce hydrogen (and oxygen) as fuel.

5. Solar-energy conversion in space for use on Earth

The maximum utilisation of solar energy can be made in an orbit around the Sun. The first step towards the fullest use of solar energy is represented by a satellite in orbit around the Earth where solar energy is available nearly 24 h a day. This approach permits solar energy conversion to be carried out where it is most effective, with only the final step arranged to take place on Earth. Similar to the contributions made by existing satellites to worldwide communication networks, power from space has the potential to provide an economically viable and environmentally and socially acceptable means to meet future world energy requirements.

Fig.1 shows the design concept for the 'satellite solar power station (SSPS)'. Photovoltaic solar-energy conversion with two symmetrically arranged solar-cell arrays represents the basic principle of the SSPS, which will be designed to produce electricity in synchronous orbit. This electricity is fed to microwave generators arranged to form an antenna located between the two arrays. The antenna directs the microwave beam to a receiving antenna on Earth, where the microwave energy is efficiently and safely converted back to electricity. In synchronous orbit, the satellite will be stationary with

Fig. 1 Design concept of a satellite solar power station (SSPS)

respect to any desired location on Earth. The use of the microwave beam allows all-weather transmission, so that full use can be made of the nearly 24 h of solar radiation available in a synchronous orbit. This availability, except for short periods near the equinoxes when the satellite enters the Earth's shadow for a maximum of 72 min a day, provides a 6- to 15-fold advantage of solar-energy conversion in space compared with terrestrial applications, where useful operations are limited by weather conditions and the day and night cycles, and installations are subjected to erosive processes and require energy-storage techniques. This advantage is translatable in terms of reduced land area and capital costs. The very high efficiency of direct microwave energy rectification into electricity in the receiving antenna reduces the waste heat generated on Earth to a fraction compared with any thermodynamic energy-conversion method.

The SSPS can be designed to generate from 3000 to 15 000 MW. Thus a network of satellites could meet a significant portion of the energy demand on a national and world scale.

5.1 Solar-energy conversion

There has been a substantial development in solar-energy conversion devices since the laboratory demonstration of silicon solar cells in 1953. Today they are a necessary part of the power system of nearly every unmanned spacecraft. The cost of these devices has decreased by nearly a factor of 10, while their efficiency has improved from 7% to about 14%. Development programmes to achieve an increase in the efficiency of silicon solar cells of about 18% over the next ten years have been outlined. Recently, gallium-arsenide solar cells have been reported to have reached an efficiency of 18%.

5.1.1 Weight reduction

A reduction in the weight of solar-cell arrays has been sought for a long time. Thus it is projected that the thickness of single-crystal silicon solar cells can be reduced to about 100 microns. The advantage of gallium-arsenide cells is that, in principle, they can be only a few microns thick, will operate at higher temperatures, and can withstand the space environment without a significant loss in their operating efficiency. Reduction in the weight of a solar-cell array can be obtained by assembling the solar cell in a blanket between thin plastic films with electrical interconnections between individual cells obtained by vacuum-depositing a metal-alloy contact material. A further reduction in weight can be obtained by using solar-energy concentrating mirrors, so that a smaller area of solar cells is required for the same electrical-power output.

Since 1965, the solar-array weight, expressed as kilogram per kilowatt, has dropped from 50 to 7. With the new blanket-type construction, this ratio is expected to reach 4 kg/kW by 1975. The use of solar concentrators is expected to reduce the array weight to about 1 kg/kW.

5.1.2 Cost reduction

To reduce costs, mass production of solar cells will be a requirement, particularly in view of the large areas that will have to be covered. Based on the experience of present manufacturers, there is a high probability that low-cost, high-volume silicon crystals of the desired thickness can be produced once a market of sufficient magnitude has been identified. Research and development efforts are underway to produce continuous ribbons of single-crystal silicon. On the basis of such mass production techniques, 1985 cost projections indicate that at least a 100-fold reduction is possible compared with present costs of solar cells (Reference 17).

Because solar cells with large areas will be required, the satellite solar-array structures can be designed to combine mechanical functions with action as a power-distribution network, to achieve high-voltage d.c. output. For the very large solar-array blankets, the solar cells can be connected in series to produce any voltage desired. At 0·5 V, for a 2 cm wide solar cell, about 50 V can be obtained per metre of solar-cell blanket. A series string of solar-cells can be assembled to build up a voltage of 50 kV or more.

Developments such as the vertically illuminated, multijunction solar cell can produce solar cells of high voltage and high efficiency. In such a solar cell, there may be 1000 junctions in series per 1 cm wide cell. Thus each cell may put out several hundred volts instead of 0·5 V.

The power busbar interconnecting the major segments of the solar cell arrays can be arranged to minimise magnetic-field interactions by the choice of power-distribution circuit arrangement. High-voltage switching circuits can be designed to control sections of the solar-cell array for maintenance and operational purposes.

5.2 Microwave generation, transmission and rectification (Reference 18)

An electromagnetic beam link can transmit power from a satellite in synchronous orbit through the ionosphere and atmosphere to the surface of the Earth. Large amounts of power can be transmitted by microwave. The efficiency of microwave-power transmission is high when the transmitting and the receiving antennas are large, excluding the efficient transmission of small amounts of power.

Ionospheric attenuation of microwaves is low for wavelengths between 3 and 30 cm when microwave power flux densities are also low, e.g. less than about 50 mW/cm^2. An efficiency of 99% is considered conservative for transmission through the ionosphere for conditions outside the range of significant nonlinear effects. Tropospheric attenuation is low for wavelengths above 10 cm, but attenuation will increase as wavelengths are reduced.

Microwave generators can convert d.c. to microwaves with high efficiencies. The use of a newly developed permanent magnet material — samarium-cobalt — will lead to substantial weight reduction. The vacuum in space obviates the glass envelope required on Earth. An efficient microwave generator can radiate waste heat directly into space by means of a heat radiator. The use of pure-metal cold cathodes will greatly increase the reliability and operational lifetime of this device.

At present, the efficiencies that have been demonstrated in the three major functions of the microwave transmission system are: 76% for the d.c. to microwave conversion, 94% for the microwave beam, and 64% for the microwave to d.c. energy conversion.

The output of an individual microwave generator weighing a fraction of a pound can range from 2 to 5 kW. A series of such generators can be combined in a subunit with individual phase control. The subunits are assembled into a slotted array-type transmitting antenna to obtain a microwave beam of a desired distribution, which can range from uniform to Gaussian. For efficient microwave transmission, the transmitting antenna should have a diameter of about 1 km. For this size antenna, the diameter of the receiving antenna on Earth would have to be about 7 km for Gaussian distribution in the beam within which 90% of the transmitted energy falls.

The receiving antenna rectifies and converts the microwave beams into d.c., which can then be fed into a high-voltage d.c. transmission system or be converted into 60 Hz a.c. Rectification can be accomplished by means of diodes, which have already been demonstrated to have a 75% conversion efficiency. With improved circuits and diodes, an efficiency of about 90% should be achievable. The resulting waste heat from inefficient rectification at the receiving site can be convected by the ambient air. The overall efficiency of microwave transmission from d.c. in the SSPS to d.c. on the ground is projected to be about 70%.

Fig. 2   SSPS dimensions

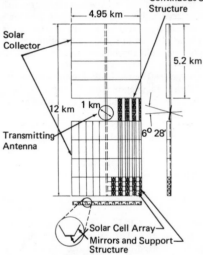

Fig. 2 indicates the major dimensions of the SSPS. Each half of the solar collector sized to generate 5000 MW on Earth is about 5 x 5 km. As these dimensions indicate, the SSPS orbital system is orders of magnitude larger than any spacecraft launched today. From an overall spacecraft design standpoint, the basic technology problems involved are related to its large size and the goal of a 30-year operating lifetime. However, the principles governing the design are founded on advances from an existing technology base.

The SSPS structure is composed of high current-carrying structural elements that will induce loads or control forces in the structure by electromagnetic effects. New design approaches will have to be developed to satisfy orientation requirements of such a large spacecraft. Flexible spacecraft structures, such as the large solar-cell array for space stations, have been studied in the past, including the maintenance of inertial pointing capabilities of structures such as a large 1000 ft-diameter antenna. Low-thrust ion-propulsion systems appear promising for flight-control purposes, particularly because of their continuous short-term impulse capabilities and lifetimes, which are compatible with the overall objective of achieving 30 years of operation.

An SSPS, capable of providing 5000 MW of power on Earth, will weigh about 25 million pounds. Such a massive satellite will require a transportation system from Earth to synchronous orbit, which will be an outgrowth of the present space-shuttle development. This transportation system will have to be designed for high-volume transport of payloads to low Earth orbit, followed by delivery of partially assembled elements to synchronous orbit for final assembly and deployment. A second-generation space shuttle, utilising returnable boosters rather than present space-shuttle solid propellants, would have inherent cost advantages. An ion-engine-propelled tug that, over a period of 6 to 12 months, would follow a spiralling trajectory to synchronous orbit, could provide an effective complement to the space shuttle. For a combined chemical/ion-propulsion system, about 500 Earth to orbit flights would be required to deliver the elements of a 25-million-pound spacecraft to synchronous orbit. Cost projections for this type of transportation system are about $100/lb to synchronous orbit.

An interesting possibility exists, of launching payloads into orbit with high-power lasers. A laser propulsion system has the advantage that almost all the equipment needed is on the ground, and does not require massive rocket engines. Laser propulsion, which would operate only a few minutes for each launch, would have an immense capacity for putting payloads into orbit. It would be possible to orbit masses that are huge compared with those at present orbited with rocket-propulsion devices. Air is used as a propellant medium; thus polluting substances would not be contaminating the atmosphere. Although the possibility of high-power laser propulsion is still largely unexplored, new possibilities are opened up, particularly for a concept such as the SSPS.

A prototype SSPS could be demonstrated by 1990, based on the necessary steps required for the development of such a system. The major tasks that would have to be accomplished are:

(a)   up to 1980: feasibility assessment, technology development, preliminary design and system verification tests

(b)   up to 1990: orbital demonstration tests of subcomponent and system functions and operational prototype system demonstrations

(c)   beyond 1990: development of an operational system.

6.   Environmental considerations

The impact of any new technology must be assessed. Such a technology assessment must include, in addition to the purely physical impact, the economic and social impacts. As an example of the assessment required, the key environmental effects of an SSPS are discussed below.

6.1   Microwave biological effects

At present, various standards for microwave exposures have been established, ranging from 10 mW/cm^2 for the USA (Reference 20) to 0·01 mW/cm^2 for the Soviet Union. The major difference in these standards can be traced to the uncertainty in the interpretation of laboratory observations of microwave exposure (Reference 21).

The US standard is based on microwave heating of body tissues. On the basis of experience with microwave equipment over several decades and the resulting exposure of a significant population to microwaves, there are remarkably few reported incidences of biological damage. Soviet investigators have indicated that the central nervous system is affected by microwaves, even at very low exposure levels. These considerations have lead the Soviet Union to set a continuous exposure standard at 0·01 mW/cm^2.

In view of this very basic difference in interpretation of the effects of microwave exposure, there is a need to develop experimental procedures, so that certain byproducts of microwave generating equipment operation, such as X rays, ozone, and oxides of nitrogen, in addition to extraneous environmental conditions imposed on laboratory test objects, would not lead to a misinterpretation of the laboratory observations.

The design approach for SSPS must recognise that a range of frequencies in microwave power flux densities may have to be accommodated, and design features will have to be adjusted accordingly. An understanding of the specific SSPS-induced environment, predictions, analyses and measurements will be an essential component of the development programme.

Precise control of the microwave beam through transmitting-antenna stabilisation and automatic phase control will ensure that the microwave power will be efficiently transmitted to the receiving antenna. The antenna size, the shape of the microwave power distribution across the antenna, and the total power transmitted will determine the level of microwave power flux densities in the beam reaching the Earth.

The SSPS design will incorporate several fail-safe features, including a self-phasing signal transmitted from the ground to the transmitting antenna, to assure precise beam pointing. Loss of acquisition of the signal will lead to demodulation of the microwave beam. In addition, remotely operated switches in the SSPS solar collector will instantaneously open-circuit the solar-cell arrays and shut off power to the microwave generators.

A wide latitude in microwave power flux densities can be obtained by selecting the transmitting and receiving-antenna diameters. The receiving antenna has to cover only an area consistent with achieving overall high efficiency of operation. A guard ring of a few kilometres can be provided, so that the level of microwave exposure outside this ring will be less than 1 μW/cm^2, which is one order of magnitude below the Soviet microwave exposure standard.

In addition to direct biological effects, interference with electronic equipment, medical instrumentation, or electroexplosive devices must be avoided. The lack of sensitivity of this equipment to a low level of microwave exposure will have to be confirmed, and, if required, industry-wide standards will have to be established.

The effects on birds flying through the beam is not known. Research on the effects of microwave on birds at the level to be encountered in the microwave beam will have to be carried out as well. Preliminary evidence indicates that birds can be affected at levels of microwave exposure of 25 − 40 mW/cm^2 of radiation in the X band. The evidence suggests an avoidance reaction by birds to the exposure.

The effects of microwave exposure on aircraft flying through the beam must also be considered. The shielding effects of the metal fuselage and the very short time of flight through the beam would not result in significant human exposure. The means for protecting aircraft fuel tanks from electrical discharges are now standard design features, but the absence of microwave-induced hazards will have to be confirmed. In addition, the extent of possible interference with aircraft communication and radar equipment will have to be established.

6.2 Interference with Radio Communication

Worldwide communications are based on internationally agreed and assigned frequencies. Because a frequency band spanning the most desirable operating frequency of the SSPS (S band) is already in heavy use, the potential for radio-frequency interference (r.f.i.) of the SSPS with existing communication systems is high. The design of the microwave generators used in the SSPS can eliminate most spurious output characteristics, but r.f.i. could occur during the shutdown of the generators as the SSPS enters the Earth shadow, or result from noise sidebands about the operating frequency or from inadequate suppression of frequency harmonics. R.F. filtering can greatly reduce these undesirable effects. However, the very large power output of the SSPS may pre-empt certain frequencies that are now in general use. The understanding of r.f.i. effects and international agreement on frequency assignments represent issues that will have to be faced at various stages during the SSPS development.

6.3 Environmental effects

The SSPS represents an approach to power generation that does not use naturally occuring energy sources, but relies on a constant and inexhaustible energy source. Thus environmental effects associated with mining, transportation, or refining of natural energy sources, are absent. Natural resources will have to be used to produce the components for the SSPS and the propellants for transportation to orbit. Nearly all the materials to be used for the components are abundant. Rare materials, such as platinum or gallium, would require less than 2% of the yearly supply available to the USA for each SSPS. Electrical power required to manufacture the various components and to launch the SSPS into orbit will be equivalent to about nine months of power generated during SSPS operation.

Environmental effects include a slight heat addition to the atmosphere due to absorption of the microwaves as the beam passes through the atmosphere. The possible effects of this heat addition on atmospheric-circulation patterns will have to be established.

Heat will also be added to the atmosphere during launch operation. Because a substantial launch frequency is required to place each SSPS in orbit, e.g. 500 launches for a 25-million-pound SSPS, local heating effects will have to be investigated, as well as the addition of products of combustion of chemical propulsion systems (primarily water vapour). If laser propulsion should prove to be feasible, the air could serve as the propellant medium, reducing the environmental impact of repeated launches. Noise pollution from the high-frequency launch operation would be of concern in the immediate vicinity of the launch facility, and would have to be reduced by suitable design techniques and the choice of a suitable location for the launch facilities.

The very high efficiency of microwave to d.c. conversion at the receiving antenna, e.g. 90%, will greatly reduce the waste-heat addition to the environment, compared with any other power-generation method based on thermodynamic principles. The heat exchange to the surrounding atmosphere from the receiving antenna element can take place by natural convection. This may lead to the formation of a 'heat island' of about the same magnitude as encountered over an urban area.

The visual impact of the large receiving antenna, e.g. 7 km in diameter, can be decreased either by suitable landscaping or by incorporating the antenna in an industrial park.

6.4 Land use

Substantial flexibility exists in choosing a suitable location for the receiving antenna. The area has to be contiguous, but need not be completely flat terrain. The location can be in a region where land is available that is not suitable for other uses, e.g. a desert, previously strip-mined land, or in an area where major electrical-power users, e.g. aluminium smelters, are located. The antenna could be located in an industrial park to serve several major users. Roofs and covered roadways could be designed to exclude microwaves from working areas completely.

The antenna structure can be designed to be mostly open, so that sunlight and rain can reach the land beneath it. Vegetation growing beneath the antenna would be effectively shielded from microwaves, and could be harvested, precluding the land from becoming a biological desert.

7. Economic considerations

Business feasibility and cost to consumers have been overriding considerations concerning energy production in the past, and it cannot be assumed that it will be otherwise in the future. However, as the population grows, pressure on resources, environmental constraints, and almost certain requirements for vast increases in the availability of electricity will make it equally unsafe to assume that established economic criteria will be enough to decide the choice of a particular power-generating system. On present estimates, a prototype SSPS will cost about 3 - 5 times more than comparable available energy-production technology based on fossil or nuclear fuel. Inspection of the components making up the cost projection indicates that major costs are contributed by the solar-cell arrays and by the Earth to orbit transportation system. There are reasonable expectations that these costs can be substantially reduced through well directed research and development programmes using approaches that, although definable, are now beyond the present state of the art.

Standards to establish the true cost for energy production will have to be developed, and hidden costs, which at present are not charged to other methods, will have to be identified, so that relative costs of other systems can be established on a comparable basis. There is as yet no consensus on the procedures that will have to be used; agreement on appropriate standards will be essential to meaningful analysis of economic impact.

The economic and technical feasibility and the environmental and social desirability of the SSPS will have to be established prior to any major commitment to the development of this alternative energy-production method based on solar energy. What is required is that sufficient information be made available so that the option represented by the SSPS can be pursued, if other approaches should appear to be less attractive, whether as the result of energy-resource, environmental, or social considerations.

8. Social considerations

To deal with the social impact of the SSPS, an assessment is required that will serve to identify all the effects of the specific technologies employed (physical, environmental, economic; direct and derivative; immediate, intermediate and long-term), so that the social desirability or undesirability of these effects can be evaluated (Reference 22).

Panels, commissions and committees have addressed the questions of social desirability over the last few years. Agreement, however, has not yet been reached on how to express 'social costs' either on the most detailed level or on an accounting basis for society as a whole. Social indicators of society's health and growth will have to be identified to determine costs of the SSPS to society as a whole, in the same way that today the gross national product is used to express the status in the field of economics (Reference 23). Appropriate standards and criteria that are developed during a process of social impact analysis will have to be tested with various groups having an interest in energy-production methods, and appropriate comparisons between solar and other energy production systems will have to be made as well.

Many groups, individuals, and sectors of society, as well as institutions, will interface in relation to SSPS. To some degree, each of these interfaces must be analysed and understood. Various communication methods will have to be employed to inform the public of the projected impact of the technologies being developed at appropriate stages of the development programme in support of the SSPS objectives.

9. Conclusion

There are several other energy sources in addition to solar energy that have the potential of meeting future energy requirements, but only very few have a limited impact on the environment and conserve the finite resources of the Earth. Solar-energy applications, represented by the SSPS, are still in an early stage of development. Feasibility assessments are proceeding on the various applications discussed here. Thus it is too early to tell which of the approaches now being studied will be judged to have the greatest potential to be of overall benefit to society. As more is learned about the operating characteristics of potentially competitive electrical-energy generating systems, the views on what 'best' performance represents will continue to evolve. Thus the criteria for decision-making — whether based on cost, resource conservation, or environmental protection — may be quite different over the next decades, and will continue to change as long as technical developments continue actively on the various energy-production systems.

Development of energy-production systems, such as the SSPS, over the next few decades will permit society to look beyond the year 2000 with the assurance that future energy requirements will be met without endangering the planet Earth. But even successful development of large-scale applications of solar energy still will require that efforts be made to reduce energy consumption and to slow growth for growth's sake.

Technology changes are creating a climate leading to institutional and social changes that, in their overall impact, can be expected to rival the 19th-century industrial revolution. Those individuals and groups that are charged by society with responsibility for leadership must face these new challenges and opportunities. Instead of creating 'dark Satanic Mills', there must be the realisation that, to survive, man must learn to 'replenish the Earth' and not just 'subdue it'.

10. Acknowledgment

The author gratefully acknowledges the encouragement and support of his associates at Arthur D. Little Inc., and the co-operation of the staff of Grumman Aerospace Corporation, Raytheon Inc., and the Spectrolab Division of Textron Inc., whose interest and commitment to the potential of power from space were essential to the development of the concept of a satellite solar power station.

11. References

1.   Electrical Research Council: 'Electrical utilities industry research and development goals through the year 2000'. ERC Publication 1-71, New York, June 1971

2.   FORRESTER, J.W.: 'World dynamics' (Wright-Allen Press, Inc. Cambridge, Mass. 1971)

3.   MISHAN, E.J.: 'Costs of economic growth' (International Publications Service, New York, 1970)

4.   ODUM, H.T.: 'Environment, power and society', (John Wiley, New York, 1970)

5.   TYBOUT, R.A. and LOF, G.O.G.: 'Solar house heating', Nat. Resources J., 1970, **10**, (2), pp. 268-326

6.   SZEGO, G.C., FOX, J.A., and EATON, J.R.: 'The energy plantation'. Proceedings of the 7th American Chemical Society intersociety energy-conversion-engineering conference, Washington, DC, 1972, pp. 1131-1134

7. HILDEBRANDT, A.F., HAAS, G.M., JENKINS, W.R., and COLACO, J.P.: 'Large-scale concentrators and conversion of solar energy', Ecology, 1972, **53**, (7), pp. 684-692

8. GLASER, P.E.: 'Power from the Sun: its future', Science, 1968, **162**, pp. 857-886

9. ESCHER, W.: 'Helios poseidon'. Escher Technology Associates, St. Johns, Mich. Feb. 1972

10. MEINEL, A.B., and MEINEL, M.P.: 'Physics looks at solar energy'. Phys. Today, 1972, **25**, (2), pp. 44-50

11. WOODALL, J.M., and HOVEL, H.J.: 'High efficiency $Ga_{1-x} Al_x$ As - Ga As solar cells', Appl. Phys. Lett. (to be published)

12. CURRIN, C.G. et al.: 'Feasibility of low cost silicon solar cells'. Presented at the 9th IEEE photovoltaic-specialists conference, Silver Spring, Maryland, 4th May 1972

13. HAMMOND, A.L.: 'Solar energy: the largest resource', Science, 1972, **117**, p. 1089

14. TABOR, H.: 'Solar collectors, selective surfaces, and heat engines'. Proc. Nat. Acad. Sci., 1961, **47**, ('Solar energy special issue), pp. 1271-78

15. MEINEL, A.B., and MEINEL, M.P.: 'Physics looks at solar energy', Phys. Today, 1972, **25**, (2), pp. 44-50.

16. ANDERSON, J.H., and ANDERSON J.H.Jun.: 'Thermal power from seawater', 1966 Mech.Engng., **88**, (4), pp. 41-46

17. CURRIN, C.G. et al.: 'Feasibility of low cost silicon solar cells'. Presented at the 9th IEEE photovoltaic-specialists conference, Silver Spring, Maryland, 4th May 1972

18. Special issue on 'Satellite solar power station and microwave transmission to Earth', J. Microwave Pwr., 1970, **5**, (4)

19. KANTROWITZ, A.: 'Propulsion to orbit by ground-based lasers', Astronaut. & Aeronaut., May 1972, p. 76

20. Temporary Consensus Standard on 'Nonionizing radiation', issued under the Occupational Safety & Health Act of 1970: Federal Register, 1971, **36**, (105), Sec. 1910.97, May 29, pp. 10 522 – 10 523

21. Special issue on biological effects of microwaves', IEEE Trans, 1971, **MTT-19**, (2)

22. Committee on Public Engineering Policy, National Academy of Engineering: 'A study of technology assessment'. Committee on Science & Astronautics, Washington, DC, July 1969

23. BAUER, R.A. (Ed.): 'Social indicators' (MIT Press, 1970)

4.2   Feasibility of fusion power

Prof. K. Husimi
Institute of Plasma Physics, Nagoya University, Nagoya, Japan.

I attended the fifth conference on Plasma Physics and Controlled Fusion Research of the Plasma Physics Division of the European Physical Society, in Grenoble. It will be appropriate to begin with my impressions of that conference.

The main topic was naturally about the up-to-date performance data of the Tokamaks of the USSR, the USA, the UK and Japan. Tokamak is a toroidal magnetic field device, designed to confine and heat hot plasma of keV orders, and it is now regarded as a single survival among a great variety of confining devices proposed at the early stages of fusion research. Originally conceived and nourished over 15 years by the Soviet Academician Arzimovitch, the Tokamak suddenly came to the focus of attention five years ago at the IAEA conference on Fusion held at Novosibirsk, where a British team confirmed the Russion results by means of laser technique. Since then, many plasma physics laboratories all over the world have regarded the Tokamak as the King's way to the future fusion reactor and have begun to construct Tokamak-like devices, more or less modified and extended from the original Russian T-3. Besides Culham laboratory of the UK, I may name Fontenay-aux-Roses of France, Garching of Germany and Frascati of Italy as having just finished the construction of their respective Tokamaks.

However, the Grenoble conference was rather excited by the dramatic appearance of a dark horse, the 'dense plasma'. The extremely intense, and at the same time coherent, light wave produced by a laser can be focused to a small target, to produce a hot plasma. This process was pursued by a second Academician, Basov, to the extent that the laser-irradiated deuterium ice could produce a small number of neutrons of thermonuclear origin. Since this is a kind of micro-hydrogen-bomb, the subsequent development was classified and remained unknown to the public in the USA. The secret was suddenly disclosed, and in the Grenoble conference Rosenbluth made a special summary report concerning the theoretical analysis of a possible micro-explosion. The idea is roughly as follows. Suppose a spherical deuterium ice be irradiated by a laser beam from every side. The surface layers will be heated and abraded, so that the remaining central core may be very much compressed. According to theoretical calculations, the compression can produce a quite dense plasma, the density of which may be a thousand times as much as the density of its solid state. Hence the name 'dense plasma'.

I do not know why the military secret was declassified at this moment of its development, but it is quite clear that we are here on the border line of peaceful and military applications of scientific knowledge. E. Teller, the so-called father of the hydrogen bomb, proudly predicts (it is reported) that within ten years a 'novel internal combustion engine' will be born based on the principle of the dense plasma.

I would like also to cite the words with which an expert in fusion at Culham laboratory answered me when I asked his opinion about the feasibility of the dense plasma concept.

Will Teller's internal combustion engine really take the place of Tokamak as the King's way to the fusion reactor? The answer was this: Everyone finds the most interest in a matter where least is known. I consider this remark can be applied to all fusion research efforts. Fusion is attractive, because it remains still in a state of theoretical possibility. The very appearance of the dark horse means that the case is not yet fully settled. This is an important point in assessing the feasibility of fusion power.

Let us now enumerate the advantages of fusion over other types of power. First, the resources of fusion fuels are virtually inexhaustible; they are so abundant that mankind can maintain the present standard of living over many millions of years. This is quite true if we prefer the deuterium-deuterium reaction. But if we are compelled to adopt the far more reactive deuterium-tritium mixture, we must take into account the scale of lithium resources. The life span is reduced considerably, to the order of a thousand years. But this is still long enough. In addition, if we include the lithium reserves in sea-water, the figure again rises considerably. In any case it is only natural to expect a rich cosmic abundance of light elements compared with heavy elements such as uranium.

To the merit of fusion, there is the even distribution of deuterium and lithium reserves when one considers the possibility of extracting these from sea-water. As history teaches us, many conflicts between nations occur because of the uneven distribution of mineral reserves. Thus one may count this point as a great merit of fusion power.

A most important point is, of course, the cleanness of fusion power. Only virtually non-radioactive materials are involved in the deuterium-helium 3(^{3}He) reaction:

$$ {}^2_1D + {}^3_2He \rightarrow {}^4_2He + {}^1_1H $$

resulting in non-radioactive products of ordinary hydrogen and ordinary helium. But many people hesitate to propose such a reaction, because it burns at a temperature about five times as high as the D-T reaction. If we are compelled to adopt the D-T reaction, we have to consider the radioactivity of tritium and the secondary activity of structural materials induced by 14 MeV neutrons. The induced activity is more or less similar to that in a fission reactor. The neutrons of the D-T reaction produce tritium in the lithium blanket, and the tritium so created should be brought back to the central hot plasma to be burned as fuel. The crucial point will be the amount of tritium radioactivity stored and circulated within the reactor. Again we can predict almost anything, since we have no definite concrete design of a fusion reactor. But one cautious analysis estimates that a fusion reactor of 1 GW output has a tritium inventory of the order of 10^8 curies, and about the same amount of tritium circulating in the blanket of lithium and in the recovery system. Although the stored activity in curies is comparable to that of a single fission-product component such as iodine-131 of a fission reactor with the same output power, the tritium emits only electrons of feeble energy, and a fusion reactor is decidedly superior in this respect.

There is also the fact that there is no possibility of explosive accidents due to runaway or excursion. However, one must be extremely cautious when handling tritium, because, in the form of tritium water, it may contaminate the whole biosphere. For this reason, a fusion reactor should not have a water cooling system, and an alkali, presumably lithium itself, molten-metal cooling system is recommended, as in the fission fast reactor. Another problem arises because hydrogen quite easily diffuses through hot metals, and it is technically difficult to stop the tritium leakage. How much tritium can be allowed to leak from the reactor complex? It is estimated that, if the 1 kw/person power is given to the global population of 6 x 10^9 people by means of fusion, then it will be acceptable to have a leakage rate of 0·14 curies/kW-year, which amounts to the 0·7 %/year leakage of inventory. This technical target requires extremely cautious design of the reactor complex, including the recovery and handling systems of tritium. But I would like to stress that there is no problem of radioactive waste disposal, because the fusion reaction does not leave any radioactive products comparable to the fission products of uranium reactors.

Since most experts in the field of fusion research are confident enough about the eventual demonstration of its scientific feasibility, and some of them are now turning to the study of the technical realisation of fusion reactors, it may be in order to point out some of the more salient points in the design of a fusion reactor. The most important feature of a fusion reactor probably lies in the low power density, low in the relative sense, of course. Whereas one deals in fission reactors with pencil-like thin fuel rods from which a huge amount of power is extracted, the same amount of power is produced within a rather large volume of hot plasma, which is in no way damaged in the usual sense because plasma can be regarded as a state of matter completely damaged. The second point I would like to note is also closely related to the low power density of a fusion reactor. About 2m thickness of moderator should receive a very high flux of fast neutrons, carrying most of the fusion power. Thus the energy deposits over a rather wide volume of moderator-absorber. Another part of the power impinges on the surface of the vacuum wall in the form of high-temperature radiation, the interaction of which should be the object of intense research.

To summarise, then, fusion power is in fact abundant and clean if special technical precautions are provided. The only drawback of fusion power is that even the scientific feasibility is not yet demonstrated, although many experts consider it realisable within five years. Thus we could say that we do not have to bother about the exhaustion of terrestrial energy sources; but the 'paradise' will come only after many years of assiduous research effort.

The last remaining problem is the thermal pollution of local rivers or sea water. This might be quite acute with fusion power since even the economically minimum size of a fusion power station requires an output of the order of several gigawatts. This can be remedied only by utilising the waste heat, for example by incorporating the fusion station in the planning of a whole city, the waste heat being used as area heating.

4.3 Longwall furnace gasification of coal underground

Dr. J.W. Taylor
Department of Fuel & Combustion Science, University of Leeds, UK

1 Introduction

Russian scientists and engineers have proved (Reference 1) over many years that it is possible to convert an underground coal seam by partial combustion into fuel gas and to lead a stream of such gas out of the ground, to burn in electric power stations for example. Basically, air is blown into the ground, and fuel gas led out of the ground from a point some tens of metres away. The gas has, however, a typically poor heating value. Experiments in other parts of the world, including some by the UK National Coal Board (Reference 2) in the 1950s, have tended to confirm the relatively poor heating value of the gas, and in consequence have shown a relatively poor economic prospect. In general the process is very wasteful, with much heat going to the adjoining rocks.

The methods that have been suggested for gasifying coal directly from a seam have followed the original basic precept of blowing in the air etc., and leaving Nature to do virtually all the rest. This precept is so attractive that it can deceive us into thinking it is the only real option. If we consider ordinary chemical plant above ground, we are accustomed to designing efficient gasification processes for such plant. In these designs we are successful because we are careful to arrange to have a close control over the variables such as position, and rate, of inflows and offtakes. If we could use this general approach underground, we could expect to avoid producing poor quality gas. In addition, we could expect a great reduction in heat losses compared with the situation hitherto, and consequently expect a considerably higher efficiency. The gasification would then be ruled by the fuel engineer rather than by Nature.

The aim here is to describe a scheme (Reference 3) by which it is hoped that this will be achieved. The underlying purpose is to reduce the overall cost of firing power stations. A further purpose is to produce a clean gaseous feedstock for synthesising petroleum and possibly substitute natural gas, having regard to the dwindling natural reserves of these compared with those of coal.

2 The longwall-furnace method (Fig 1)

The method may be explained as follows:

(i) Simple machine: To achieve the necessary close control over the gasification of a coal seam, a very long but relatively simple machine would be placed against the exposed coal seam (the longwall face). The diagram shows a (purely schematic) vertical section of this arrangement.

(ii) Constant-width 'duct': This machine might, in principle, be said to function as a horizontal producer-gas/water-gas generator consisting of a very long 'duct', one side of which is formed by the coal face, and whose width is kept substantially constant as the coal is consumed. The oxidant gases: air (oxygen), steam etc. would be fed in at suitable

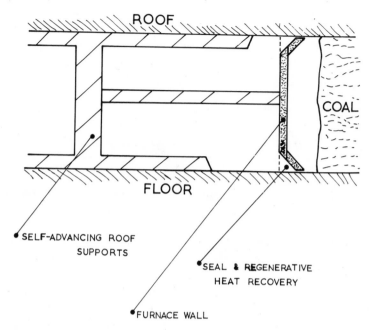

Fig. 1   Longwall furnace, vertical section (purely schematic)

points along this 'duct' in a swirling motion, and the product fuel gas taken off at a
point sufficiently far along the coal face for the gas to have an optimum composition.
(iii) Entire coal face length: Beyond this point, a similar arrangement of gas flows
would constitute a second gasification unit, and then probably a third etc., extending
the full length of the face.
(iv) Creeps forward: The 'duct', or furnace space, would be formed by a slowly moving
furnace wall positioned parallel to the coal face and provided with suitable sealing
members against the roof stratum and the floor stratum, respectively.
(v) Remotely-operated: The furnace wall would be held by a self-advancing roof-support
system, of a kind much as in British mines at the present time, and this complete machine
would be moved forward by remote operation.
(vi) Ash stays underground: Ash or slag would be left behind, (probably by) being over-
ridden as the furnace advanced.
(vii) Safety and thermal efficiency: Parts of the inflow gases would cool the roof and
the floor before these became exposed to the ambient atmosphere of the mine.
(viii) Only pipelines needed: Pipelines would take the gas at about ambient pressure from
the coal face to the pithead, and from thence (possibly along the side of existing coal-
supply railway lines) directly to the power station. There it could be fed into the boiler
without the handling costs associated with solid coal.
(ix) Cleaner mines and power stations: Winning the energy of coal would no longer involve
the production of coal dust with its associated explosion and health hazards. At power
stations there would likewise be no coal dust, nor ash to deposit throughout the flow
system or to dirty the whole boiler area.
(x)   Easy sulphur removal: Gasification is a preferred method of removing sulphur from
coal. Being produced as a gas, the fuel from the longwall furnace would already be in
the form in which it is relatively easy to remove polluting sulphur compounds. This is
in sharp contrast to removal from solid fuel or from flue gases.

3 Advantages over previous approaches to underground gasification

Rock falls into the combustion region would no longer be an inherent part of the process, thus avoiding the possibility of extensive loss of gas through cracks in the strata. The geometry of the gasification channel would no longer be subject to enormous variation since the channel could be maintained at an optimum width. Combustion of the fuel gas with bypassing streams of air would thus no longer be a possible cause of the lowering of the heating value of the final product.

The fuel gas offtake point could be chosen to avoid both early offtake, which leaves too much useless diluent (carbon dioxide) gas in the fuel, and late offtake, which is a waste of productivity. Steam injection is a vital part of efficient gasification above ground. The longwall-furnace method would allow the entry points to be chosen (typically in high-temperature regions) to give the maximum gasification efficiency. In general, because of the possibility of close control, the various inflows could be 'tailored' to optimise the reaction.

By the use of heat-exchange and regeneration principles, heat could be conserved much more effectively than in conventional underground gasification. Apart from leading to higher efficiency, this would facilitate the maintenance of higher reaction temperatures, and hence the attainment of a more complete chemical reaction.

4 Economic advantages over solid coal for power stations

In the UK, as in many parts of the world, we are anticipating a continuing enormous increase in the demand for electricity. We need a low-cost fuel to make the electricity. As is well known, handling a solid is always more expensive than handling a fluid material. Thus a gaseous fuel should be cheaper to bring to the surface, cheaper to deliver to power stations, and considerably cheaper to handle at power stations, than solid coal. The truth of this is made clearer when we realise that by using pipelines we could dispense with whole complexes of equipment. These include conveyors, wagons and wagon lifts in the pit, screening at the pithead, the entire (railway) transport facilities for supply to power stations, and at the latter we would no longer need wagon tippling, conveyors, bunkers, or grinding mills. Possibly most important of all, we would no longer need the very expensive ash removal and disposal operations at the power station.

5 Feedstock for synthesis of petroleum or substitute natural gas

World reserves of natural gas and petroleum are much less than those of coal. They are to be measured in terms of decades rather than of centuries of consumption. These facts have become more widely recognised in the last few years. It has become clear that if we are to have the convenience of liquid and gaseous fuels, mankind will have to synthesise these fuels increasingly from coal. Natural gas resources are likely to be the first to constitute a serious problem. This is particularly true of North America. It is necessary to find means of synthesising substitute natural gas (SNG) to supplement the gas from wells when this begins to be in short supply. To synthesise this from coal in preference to petroleum is clearly consistent with long-term needs, and with short-term needs if security of supply is taken into account.

Apart, therefore, from the possibility of the longwall-furnace method being used to supply power stations, we should consider it as a possible means of providing a convenient feedstock from which to synthesise petroleum and SNG. The longwall-furnace gas would be expected to be suitable for use in the synthesis of the former, and possibly of the latter. The advantage of this approach would be that the input cost of the feedstock could well be less than that of solid coal. In any event, the processing would be significantly cheaper and cleaner with the gaseous feedstock than with the solid one. One might envisage a continuous supply of gas from several longwall faces, feeding a petroleum or SNG synthesis plant near the pithead, and so yielding a continuous output of these fuels.

North Sea gas supplies appear to be in abundance at present. We in the UK are fortunate in this and in the fact that oil is also being found. However, it would be prudent to regard this time of abundance as a breathing space and to use it wisely, so that in perhaps a decade or so, when the prospect has become less rosy, we shall be fully ready with satisfactory alternatives.

6 Problems which warrant attention

Development of the longwall-furnace scheme would offer some challenging problems to both engineers and research workers. Notable among these would be the questions:

(a)   whether the gasification could be made to proceed fast enough to justify the capital and running costs;

(b)   whether the gasification really could be contained satisfactorily in the region of the coal face by a suitable system of seals;

(c)   whether excessive deposits of slag or ash in the gasifier could be avoided;

(d)   whether the role of the currently used packs of rock, which control the final collapse of the roof, could be provided for in some remotely operated way.

In view of the above mentioned advantages of the longwall-furnace scheme, it would seem to be well worth while trying seriously to find ways of overcoming or resolving these and related problems.

7 References

1   KREININ, E., and REVVA, M.: 'Podzemnaya gazifikatsiya uglei', Kh. Izd. Kemerovo, 1966

2   'The underground gasification of coal', National Coal Board (Pitman, 1964)

3   TAYLOR, J.W., Brit. Pat. 1 120 384. Application date 2nd Feb 1966 (Application date in USA 8 June 1967)

4.4   Mole mining

Prof. M.W. Thring
Queen Mary College, London, UK

1   Statement of the problem

It is quite clear that the world's coal resources are substantially more than ten times as great as the resources of oil and natural gas put together. This therefore represents a vast source of energy that we have to use at least in the interim period until we have developed a completely self-renewing solar energy source. Now it is also quite clear that, if, as they have done, the Russians could operate a telechiric machine on the moon to collect samples and do other tasks up there, controlled by radio from men on the ground, we can certainly develop machines that will do all the required mining operations under the earth without men going down below the surface at all. It is equally true that, as the more accessible coal resources are used up, it will become necessary to use more inaccessible coal, for example, in deeper seams, in seams only a few centimetres thick and seams far under the sea. These are all situations where it becomes increasingly dangerous to send men to work: moreover, men must return to the surface after an 8-hour shift so that they spend a great deal of time travelling to and from their work at the coal face. In fact, one could go so far as to say that the principle reason a ton of coal on the surface is much more expensive than the equivalent amount of oil is that oil can be pumped out of the ground, whereas coal requires a mine in which men work. Such a mine must be made safe, it must be made accessible for the daily journey, it must be made suitable for men to work in with regard to comfort, height and convenience, and it must be ventilated clear of methane and also pumped dry of water. Even when great efforts are made, there are still mining disasters in which people are killed.

All these problems can be overcome if we design mole machines that will do the mining operations controlled by three shifts of men on the surface and be designed to work under these conditions — such machines can work under water or in badly ventilated conditions, and will not have to come to the surface for daily refreshment.

One way of doing this is to gasify the coal underground and turn it into a low-grade producer gas and then bring this producer gas to the surface where it can be burned in a boiler to make electricity. This process of underground gasification has been studied in Britain, Russia, America, Belgium and other places. It has many disadvantages, because it requires men to go down and prepare the system and then the machine to set fire to the seam, and it has also been shown, by some calculations that I did from the published gas analysis 30 years ago, that most of these processes leave the bulk of the coal underground as a partially carbonised coke and the gas produced consists mainly of the volatiles drawn off by the heat of partial combustion. It is also dangerous if done in the neighbourhood of other workings, in that the ground is liable to be full of cracks and a lot of the gas will be lost through them instead of coming back to the surface.

Another alternative that I and others have examined is the possibility of burning the coal completely with a small high-pressure combustion region in front of a self-propelling mole, which is a gas turbine with an electric generator. In this case, by passing air down to the system from the surface, the coal would be burnt in front of the machine after compressing the air, the high-pressure combustion gases would pass out and expand through the turbine, which drives both the compressor and the generator, and the waste gases would escape to the surface and the electricity brought to the surface. There are many problems with this system, including the problem of preventing the turbine being choked with ash, and sealing the machine so that a pressure of 10 atm. can be in front of it in the combustion region while atmospheric pressure is behind it.

However, there are so many cases where it is necessary to bring the solid coal to the surface — for example, if it is required to carry the solid coal many miles from the mine, or to use it for iron making — that I shall here try to work out the development of a mole miner that could crush the coal and bring it to the surface. It is then available at the surface for making into coke for blast furnaces, for firing in electric generator boilers, or in steam-raising boilers and for furnace heating, or it can be piped in water for considerable distances.

Just before the Second World War, there was developed an extensive system of carrying pulverised coal in lorries and delivering it to large central-heating plants and similar installations. This was stopped by the arrival of oil, which is a considerably more convenient fuel and has less ash problems, but systems of this type might have to come back when oil supplies become short again and we have to make use of our coal. If such systems were used in association with district-heating schemes, the difficulties of distribution and ash disposal would be relatively easy to overcome because of the larger scale.

2 Mining strategy

It is naturally not possible to say exactly how the system would work when the necessary ten years of research and development have been put into it. Moreover, it is probably desirable to develop a mole system that will deal with metal ores and other valuable materials, even search for diamonds or gold in a vast mass of rock, and therefore much of the research done for the coal mole will also be applicable to these other machines. Many of the early experiments will be done using existing pits to develop the first machines. However, I propose to give the scenario for two possible ways of working such a system when it has been finally worked out, just to demonstrate its practicability.

The simplest way of demonstrating this would be to say that one could use the identical methods of mining as at present but with all the men underground replaced by telechiric machines controlled by men on the surface. Thus the detailed operations of studying the seam, machine handling, maintenance and repair of machinery, installation of additional machinery and so on would all be done by human hands controlling mechanical hands at the coal face, with the necessary feedback of sight and other senses and using the same tools and instruments as the men would use if they were there. However, this is obviously a waste of the possibilities of machines to make mining very much simpler and cheaper, and I will therefore try to work out two possible ways of doing it, taking full advantage of the possibilities of such machines.

3 Tube-mining system

In this system, a single mole miner is connected to the surface by two flexible tubes and cables which it drags behind it as it crushes a circular hole through the coal seam, the diameter of the hole being approximately equal to the thickness of the seam. When it has proceeded the full length of one hole, it returns to the main shaft and bores another hole, alongside the first one but separated sufficiently from it so that the ground does not collapse. This is shown diagrammatically in Fig.1. It crushes the coal finely, and pumps it to the surface in water in one tube, the water being brought down in another tube. The air and liquid fuel (or AC) necessary to operate the machine are brought down

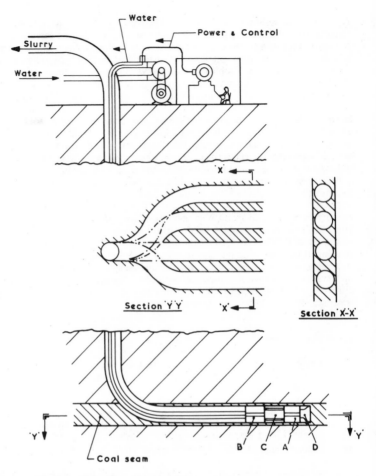

Fig. 1 Operation of mole miner

in separate tubes, and a coaxial cable carries the control signals that enable the man on the surface to steer the machine to remain in the seam when its own sensors detect a change in direction or a fault in the seam. The coal-and-water slurry is pumped directly to a power station, and burnt without separation so that no man ever sees the coal. One power station could be operated by ten or so of these miners.

This machine can crush its way through hard or soft rock and go round corners, dragging two connective flexible tubes behind it through the hole it has made. Water will be pumped down one tube and the machine will pump the water with the crushed rock products up the other. This machine will be capable of being steered from the surface to cut round a corner in quite a small radius. It will consist of two parts, as shown in Fig.1. The front part A will have three feet on it, and will carry the motor driving the cutting head D, which will have a combined hammer and drill movement. In use, the feet on the rear part will be jammed in the hole while the three rams C push the front part forward for the whole length of the stroke during the cutting action. By working these three rams unequally, it can be made to cut round a corner. When the full stroke forward is reached, the feet on the front portion are pushed out to jam this portion in the hole and then the feet on the rear portion are withdrawn and the rear portion B is brought up to the front portion by reversing the stroke of the three main rams. The rear portion drags with it the pipe and all the other connections to the surface. The cutting head will measure directly the hardness of the rock at the top and bottom and on the two sides of the hole, so that the man on the surface knows when the machine is working at the boundary between, for example, soft coal and hard rock. Measurements will also be made of the density of the rock by means of gamma rays, and possibly other physical and chemical characteristics of the four samples would also be continuously observed and the result transmitted to the man at the surface. It is even possible to have a small stream of oxygen going down a very fine pipe and to burn the coal samples and measure their calorific value. Probably the cutting head will be driven by a very-high-pressure free piston hammer with a single-cylinder liquid-fuel-injection 2-stroke diesel cycle burning with compressed air or oxygen. In this case, the combustion products will be used to aerate the rising stream and reduce the density, since the exhaust pressure will be equal to the prevailing hydrostatic pressure.

In working a coal mine, such a machine could be arranged to dig its way down from the surface to an appropriate point in the seam, and then to work outwards from this point by cutting two or three cuts, say north and south, and then running a series of cuts east and west from the first one moving a distance of up to half a mile from the main gallery in each case. It is also possible to make it so that it will cut away to the given distance, and the front head will then turn round and cut back alongside the body, which returns in its original hole. In this case, it is possible to arrange to feed washery material, or other incombustible material, down with the water and fill up the hole left behind the machine on its return. One would probably not attempt to remove all the coal but only about half of it, leaving pillars between the successive holes so that there would not be surface subsidence as in the present long-wall type of mining. The mining would be more equivalent to the original bore and pillar method, but with the galleries only the diameter of the machine, which might be half a metre, although it would preferably be equal to the thickness of the seam. If the machine indicated that it was suddenly cutting with all four edges of its cutter into rock, the men on the surface would know that there was a fault there and would have either to decide to return it to the gallery, cutting back alongside, or, if they knew the exact details of the fault, they could tell it to work its way up or down to the next part of the same seam. The machine could alternatively work out from existing coal-mine shafts or from a position where the seam outcrops.

4   The long-wall mining process   (Fig.2)

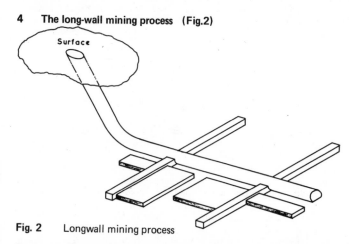

Fig. 2   Longwall mining process

To describe this process, it would be necessary first to describe six machines that would have to be specially developed. Only one of these machines has to be in constant communication with its controlling human on the surface — that is, it is a true telechiric miner. This one is described first.

4.1 Machine 1: Telechiric miner and haulier   (Fig.3)

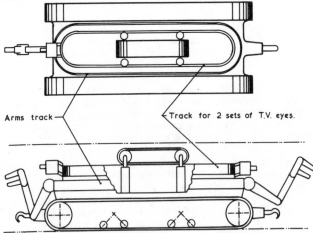

Arms track—

←Track for 2 sets of T.V. eyes.

Fig. 3

There may be ten or so of these machines working down a given mine, perhaps one to each long-wall working face. They come to the surface only for annual servicing, as they are fed with their energy by electric cable from the surface and they have sufficient battery capacity to run a mile or so out of contact with the cable. Each one is also in constant microwave communication with the surface, either by a coaxial cable or by radiowaves transmitted through the open space of the mine tubes. This communication is 2-way, so that the machine can send information, e.g. a t.v. picture of what it sees, or the measurements it takes, to its controlling man on the surface. Equally, he can give it instructions to move in any direction and to operate tools that it carries with its giant-strength arms and hands as if he were present at the spot. It propels itself by two tracks, one underneath it and one on the top that can be forced up by a vein to jam against the upper surface of the shaft. By propelling these two tracks, it can therefore move in a shaft of various diameters and at various angles to the horizontal right up to the vertical, and has a very powerful drive motor so that it can pull great loads through a mining system. It also carries a hydraulic actuator and a store of hydraulic fluid that it can replenish from a pipe to which it can plug itself so that it can operate a number of hydraulic machines. It has batteries so that it can run for 1 hour severed from its connections, and a small computer and tape recorder for instructions so that it can continue to carry these out.

4.2 Machine 2: Roadway cutter and reinforcer (Fig.4)

This machine can cut a roadway with a flat floor, circular sides and an arched roof through the various rock strata and the seam. This roadway can vary from a width equal to the height of the arch, which is one and a half times the thickness of the seam cut, to three times the height of the arch; the shapes of roadway are shown in Fig.5. It carries its own power cable from the surface, and is permanently connected to this; it is also fed with ready-mixed concrete through a tube from the surface or from a rail wagon and with a continuous steel strip unwound from a drum in a helix all round, so that it can lay a reinforced concrete arch around the whole of the roadway as it cuts it. It would cut rings E-E every so often which are deeper into the arch and into the ground and thus make a thick reinforcement for the supporting arch. It can also leave holes in the side of the roadway for the entrance of the machines into the seam. It might, for example, cut a 45° sloping roadway down from the surface to the level of the seam and then cut a main roadway along the seam with branch roadways every kilometre. From each of these branch roadways, the long-wall working systems would go out blind-ended on each side. The spoil of roadways would be carried to the surface by machine 6 (Section 4.6) operating in the roadway behind it.

170

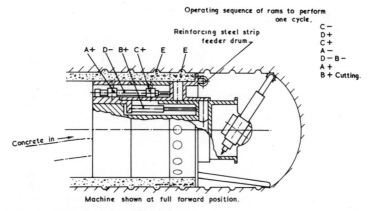

Reinforcing steel strip
feeder drum

A+ D− B+ C+   E   E

Concrete in

Machine shown at full forward position.

Fig. 4   Roadway cutter and reinforcer

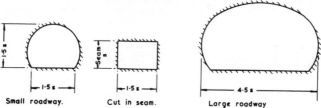

Small roadway.   Cut in seam.   Large roadway

Fig. 5   Shapes of roadway

4.3 Machine 3: Coal cutter (Fig.6)

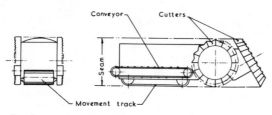

Conveyor   Cutters

Seam

Movement track

Fig. 6

This machine looks like existing ranging drum shearers, except that it has an extra cutter so that it can cut a square hole in the direction in which it moves into the coal seam. It can propel itself in and out of the first hole blind-ended. When it reaches the required distance down the hole, dragging its coal transporters behind it, props having been placed behind it by a 'miner', it then switches over to the side cutter, which cuts to the side of itself and then cuts back to the roadway along a second hole adjacent to the first one.

4.4 Machine 4: Mobile prop (Fig.7)

This consists essentially of four hydraulic jacks, each with a support plate on the top. There are two pairs that support the starting side of the mining tunnel and two pairs that can be brought close to the first pair or pushed by means of horizontal telescopic rams to a distance equal to up to twice the width of the mining tunnel. The two rams on the mobile side can also be turned down to a horizontal position, so that the machine can work under cables and supports, and then they can be lifted again. These machines

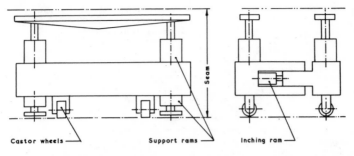

Fig. 7 Mobile prop

can therefore be moved one mining-shaft width sideways while still supporting the roof, by the operation of lowering the two on the side in the direction to which it is to be moved until they are out of contact with the roof, then extending the horizontal rams to move these forward to twice the distance, raising these rams into contact with the roof, lowering the two rear rams and returning the horizontal rams to one width again, dragging the two rear ones across.

4.5 Machine 5: Coal moving device for mine shaft (Fig.8)

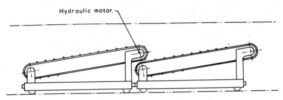

Fig. 8

This consists of a whole series of units that are placed end to end, and each of them has a belt carrying shifter bars that can be rotated by its own motor from the electricity supply. Each belt has one shifter bar every two feet in the coal within that belt, and thus the lower half of the belt is pushing the coal along that section slightly uphill. At the end of the section, the coal falls down to the beginning of the next section; thus, by driving all the sections separately, the coal is moved continuously from one end to the other. On each section, the coal-moving bars enter the coal stream as it is falling down the slope at the end, so that it is fairly well spaced apart. The series of coal conveyors are placed inside the supports all along the mining shaft, and bring the coal out to the main roadway, where it is loaded into machine 6 (Section 4.6).

4.6 Machine 6: Roadway haulier (Fig.9)

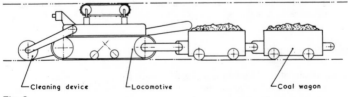

Fig. 9

This machine has not only to carry coal from the various roadways up the sloping pit shaft and to the surface, but also to bring down various requirements from the surface to the coal face and to transport coal machines to the surface when they need to be given a greater repair than the telechiric miners can do on the spot. It would probably therefore consist of an electrically driven locomotive that could tow a series of small wagons; these wagons would have broad flat wheels running on the surface of the mine, which would be kept swept by a cleaning device in front of the locomotive. All these locomotives would be controlled by modulating the main power stream from the surface, so that they could be operated not to run into each other, to go round corners and to stop at the required places.

With these machines, it would be possible to dig and operate a complete long-wall mine to remove all the coal from a seam over an area of several square kilometres without any man ever going underground at all. The mine would normally operate full of air, but it would not matter if there was a significant amount of methane in the air, and it would only be necessary to drain the workings of the mines so that the water level did not rise above a few centimetres. In the event of an accident, at the worst a few machines costing £100 000 each would be lost.

4.5   External combustion engines

S.S. Wilson
University of Oxford, UK.

The limitation of the internal-combustion engine for road-transport purposes are becoming increasingly evident, so that attention is turning to alternative prime movers. These could mostly be described as external-combustion engines, but the true difference is that they are continuous-combustion rather than intermittent-combustion engines. It is the intermittent nature of the combustion process in both spark-ignition (petrol or gasoline) engines and in compression-ignition (diesel) engines, which is the root cause of both atmospheric and noise pollution.

In the short time available for the initiation and completion of combustion, perhaps only 1 ms, it is impossible to control the combustion processes sufficiently well to ensure complete combustion (to avoid the presence of unburnt hydrocarbons or carbon monoxide in the exhaust), to avoid local overheating leading to oxides of nitrogen, or to avoid the sudden rises of pressure which lead to the characteristic knocking sound associated particularly with diesel engines even under normal conditions and with spark-ignition engines under abnormal conditions. The noise problem of a diesel engine is probably a more intractable one than the emission problem, though the noise can be reduced by proper design and acoustic treatment — at a cost that operators and the public are not yet prepared to pay.

Continuous combustion, on the other hand, whether internal as in the combustion chamber of a normal gas turbine, or external as in a steam engine boiler or in a Stirling engine, offers the possibility both of quietness and of a fully controlled, and therefore minimum-pollution, process. Careful design and control is still needed, to avoid excessive temperatures leading to the production of oxides of nitrogen, and the overall efficiency of the engine is also important, since otherwise a less efficient engine must burn more fuel for the same output, leading to an increase in exhaust gases, such as carbon dioxide, and to greater depletion of energy resources.

In approaching the problem of designing or evaluating rival prime movers, one can consider it as a problem first of all of matching and then of optimisation. Fig. 1 shows diagrammatically the five elements and the four separate matching problems. The first match is between the source of heat and the working cycle; this is a problem ignored by text books, which offer only the Carnot cycle as the ideal cycle to be aimed at, embodying isothermal heat reception at a high and uniform temperature. Unfortunately, virtually all real sources of heat are finite, so that, as heat is transferred from them, their temperature falls considerably, so the Carnot ideal is impracticable and misleading. It is significant that all real heat engines except the Stirling engine approximate to cycles in which there is a large variation of temperature during heat reception — the Rankine cycle (for steam and other vapour-power units), the Otto cycle (for internal-combustion engines) and the Joule or Brayton cycle (for gas turbines). The attempt by the Stirling-engine designers to achieve isothermal heat reception leads to the major difficulties in designing the various heat exchangers that account in large part for the lack of commercial success of this type of engine, despite the efforts of the Philips Co., and others for over 30 years.

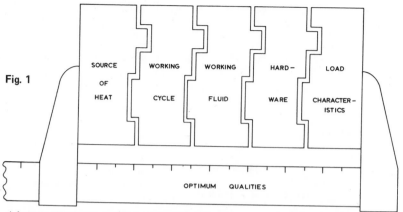

Fig. 1

SOURCE OF HEAT | WORKING CYCLE | WORKING FLUID | HARD-WARE | LOAD CHARACTER-ISTICS

OPTIMUM QUALITIES

A better criterion for yielding the worth of a cycle is the ideal trilateral cycle, made up of three lines on the temperature/entropy diagram — the curved heat-reception process (matched to the source of heat), the isentropic expansion process forming a vertical straight line, and the isothermal heat-rejection process forming a horizontal straight line. An isothermal cooling process is realistic as an ideal, since the surrounding air or water provides an almost unlimited heat sink whose temperature changes but little.

The second match concerns the working cycle and the working fluid; most engines use either air (IC engines and gas turbines) or steam, but many other fluids exist that, in particular cases, offer a better alternative. Thus the 'Freon' refrigerants have temperature/entropy diagrams with a saturated-vapour line that is nearly vertical, so allowing a cycle to be constructed that approximates to the ideal trilateral cycle. Fluids of even heavier molecular weight have saturated-vapour lines with a pronounced positive shape. This implies that the vapour after expansion is superheated (rather than wet as with steam), and the cycle efficiency would be low without the use of a heat exchanger that transfers the heat from desuperheating the exhaust, to heating the liquid being fed back to the boiler from the condenser. Thus the match between cycle and fluid can be improved.

The third match is between fluid and hardware; the available forms of expander and compressor are shown in Table 1, and the inherent characteristics of each main type of machine are shown at the foot of each column. Briefly, the reciprocating machine is a high-pressure, low-flow machine, whereas the turbomachine has low pressure ratio per stage and high volume flow. The rotary positive displacement machine (e.g. Roots blower or Wankel engine) has characteristics between the other two types. The optimum match depends on choice of fluid, cycle conditions, size of unit (i.e. power), and on rotational speed. Thus, for steam at low powers and speeds, a reciprocator is the best choice — hence the work on developing steam cars — but at large powers and speeds the turbine is to be preferred. However, with a heavy-molecular weight fluid, it is possible to use a turbine efficiently right down to powers of 1 kW or so — hence the work on vapour turbines for city vehicles etc.

The last match concerns the relation between the hardware and the load characteristics, e.g. the required torque/speed curve. For a vehicle, the most desirable curve is one having a high torque at starting, diminishing in a hyperbolic curve as the speed increases to its maximum, ie. a constant power curve. The usual internal-combustion engine has quite the wrong basic characteristic, having no torque at standstill and a roughly level torque/speed curve. Because of these deficiencies, a clutch and multispeed gearbox or an automatic hydraulic or electric transmission must be used. In comparing different systems, the transmission system required must be weighed in the balance as well as the engine. A further consideration is the ability or otherwise to store energy to meet the sudden power demands followed by idling that are so characteristic of driving, specially in cities. Because there is no possibility of energy storage with internal combustion engines, the usual solution is to put in an engine that is far too big, just for occasional full-power use. The rest of the time the engine is running at part load, at well below its best efficiency, and, even at standstill, the engine continues to consume an appreciable quantity of fuel.

Table 1

Heat-engine types

	Positive displacement		Turbomachines (Rotodynamic)	Fluid dynamic
	Reciprocating	Rotary		
Pumps	Piston Ram Diaphragm	Gear Vane Screw	Centrifugal Axial Mixed-flow	Electromagnetic liquid — metal pumps Jet pumps
Compressors and vacuum pumps	Single and multi-stage piston machines Diaphragm compressors	Roots blowers Eccentric-vane compressors Helical-lobe (screw) compressors	Centrifugal compressors Axial com-pressors Fans	Ejectors
Vapour-cycle expanders	Single and multi-stage steam engines	Various un-successful 19th — century designs	Steam turbines	Nozzles
Internal-combustion systems	Spark ignition and compression-ignition engines	Wankel and other rotary engines	Gas turbines	Rockets Ram jets Pulse jets M.H.D.
Characteristics	Unbalanced inertia forces $a\,N^2$, hence slow speeds; cool-ing possible. High pressure ratio, low throughput. Rubbing surfaces — wear, lubrication problems. Upper size limited.	Better balance, so higher speeds, more throughput, less cooling possible. Pressure ratio limited. Rubbing or leakage.	Very high speed. No cooling possible. Low pressure ratios per stage, high throughput. No mechanical wear. Lower size limited.	No good for shaft power directly, but thrust or electric power possible. Restricted field of application.

Now it is not any individual match that matters so much as the performance of the whole outfit; this is indicated in Fig. 1 by the gauge marked 'optimum qualities', that which is to be optimised is no such simple factor as thermal efficiency, but rather a whole range of factors (Table 2), tangible and intangible, so that no simple optimisation is in fact possible, but the final choice becomes a matter of compromise, usually between conflicting factors.

Table 2

Factors affecting choice of prime mover

1   Overall costs, made up of:

(a)   capital cost, interest rate

(b)   depreciation, expected life

(c)   running costs — fuel and supplies, including transport costs, duty and taxation

                — wages and salaries

                — maintenance and repairs

(d)   transmission and distribution costs

(e)   load factor — ratio $\dfrac{\text{average load}}{\text{system capacity}}$

(f)   reliability — affects provision of spare capacity

(g)   cost of cooling, or value of waste heat or other products, e.g. fly ash

2   Pollution costs — atmosphere, acoustic, thermal, visual

3   Size and weight — affects costs

4   Form of output, e.g. torque/speed curve

5   Thermal efficiency, specially in its effect in reducing size and cost, as well as fuel consumption

6   Controllability — response time, ease of remote control, part-load performance

7   Intangibles — tradition, prestige, prejudice, fashion, novelty

The process of matching may be illustrated by Fig.2, which shows curves representing the five elements; Fig. 3 shows one obvious optimisation that is usually considered the most important — lowest economic costs, as defined in purely money terms. On the other hand, Fig. 4 shows a quite different kind of optimisation — the health and social costs represented by illness, injury and death.

A recent review of alternative power plants for automotive purposes was given by E.M. Estes at a meeting on 13 March 1972 of the Institution of Mechanical Engineers in the UK. Mr Estes is a Group Vice-President of General Motors Corporation, New York, and he considers that sufficient improvements can be made to internal-combustion engines to remain acceptable. But he admits the gas turbine as a likely contender, specially for large commercial vehicles (260 kW or so), less likely for private cars because of generally poorer performance relative to conventional IC engines — poorer in acceleration, flexibility and fuel consumption. Though emissions generally are better, the nitrogen-oxides level would still not meet the 1975-76 US standard.

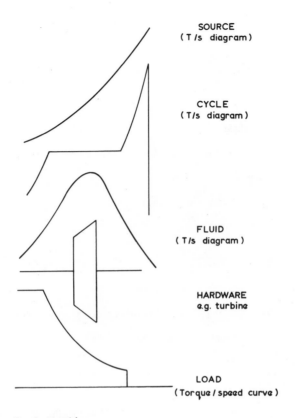

SOURCE
(T /s diagram)

CYCLE
(T/s diagram)

FLUID
(T /s diagram)

HARDWARE
e.g. turbine

LOAD
(Torque / speed curve)

Fig. 2 Matching

Though battery-driven vehicles cannot strictly be considered in the external-combustion category, in effect they are, since they rely for their energy on electricity generated by steam turbines, using either fossil or nuclear fuel.

Hence the problem of pollution is removed from the streets to the power stations, but cannot be regarded as eliminated. An appreciable night load spread over 8 h for charging purposes would be useful in raising the average load factor, but any proposal for quick charging during the day would be most unwelcome and impracticable. Though electric vehicles have been in regular use in the UK for duties such as milk delivery, they have severe limitations for more widespread use, e.g. as city passenger cars. With the usual lead-acid batteries, the performance as regards speed, acceleration and range is limited, mainly because roughly one-third of the total weight is battery weight, one-third vehicle and only one-third payload. Despite intense search for an alternative to the lead-acid cell, no viable solution has appeared; so the future of the all-electric vehicle appears to be limited to city vehicles with a low range requirement, though they can still make a significant contribution to the urban transport system.

Another means whereby batteries can help is in the so-called hybrid vehicle, in which the battery provides energy storage to even out the power demands on the engine between the peaks, for starting, acceleration and hill climbing, and the troughs, when no power is needed at standstill. By such means, the power and size of the engine itself can be much reduced, so that it can be designed to run at optimum conditions with minimum pollution products; even a spark-ignition engine could be acceptable in this form, but almost any other prime mover could be considered — Stirling engine,

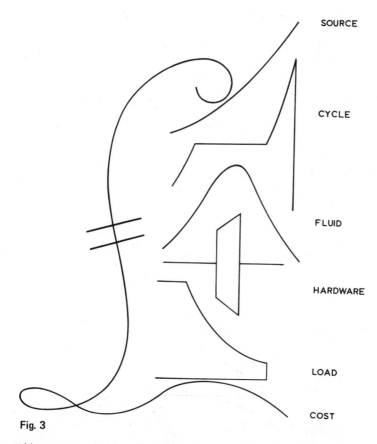

SOURCE

CYCLE

FLUID

HARDWARE

LOAD

COST

Fig. 3

gas turbine or vapour turbine. The engine power needed is determined primarily by all-up weight and by the maximum speed required; there is a big increase in power with increasing speed; e.g. for a medium-sized American car of 4000 lb all-up weight, 55 hp is needed at the wheels for a speed of 80 miles/h on the level but only half that at 40 miles/h. Since this lower speed is adequate for city driving, provided that there is adequate acceleration and hill-climbing ability, a hybrid vehicle is attractive, particularly as, with electric final drive, the right sort of torque/speed curve is available without a change-speed gearbox and clutch, so that a fully automatic transmission system is inherent in the system.

With regard to the choice of prime mover, an internal-combustion engine has advantages, but existing engines would not be suitable as they stand, since they are not designed for continuous duty and would be too big. General Motors has built a small 2-seater car with a gasoline engine of only 12 in^3, compared with 230 in^3 for a medium-sized 6-cylinder vehicle. It gave a top speed of 35 miles/h and a fuel consumption of 50 miles per US gallon.

They also built a 4-seater vehicle with a Stirling engine of 8 hp having a top speed of 60 miles/h and a fuel consumption at 30 miles/h of 30 miles per US gallon. Neither of these vehicles was regarded as an economic proposition, though, no doubt, with development, they could be reduced in cost and improved in performance. They would, of course, benefit greatly from a better battery than the lead-acid cell.

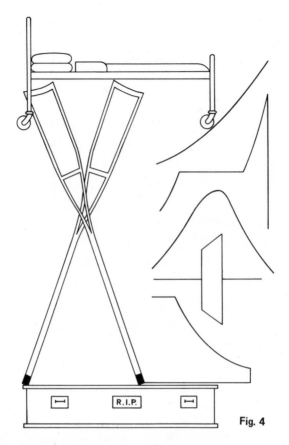

Fig. 4

The Stirling engine has many attractive features — quietness, low noise and vibration, and low emissions except for nitrogen oxide. However, it is heavy, complex and expensive; many of these difficulties can be traced to the fundamental weakness previously mentioned — the misguided attempt to achieve isothermal heat reception; the result is a complex of sophisticated heat exchangers, using expensive materials in assemblies that are difficult to manufacture commercially, while, for good efficiency, a high maximum temperature of combustion products must be achieved, leading to the formation of oxides of nitrogen.

Finally, we come to vapour cycles, of which there are two main contenders, the steam reciprocating engine (though other forms of expander are possible) and the 'organic' turbine using a working fluid of heavy molecular weight. Steam, which has a molecular weight of only 18, has a high latent heat of vaporisation, so only a small mass flow is needed in a vehicle engine; and since, in a vehicle, condensation must take place at a temperature of about 100°C, corresponding to an exhaust pressure of near atmospheric, the volume flow at exhaust from the expander is small, and a reciprocating engine is appropriate. However, since the boiler pressure is likely to be high, perhaps 1000-1500 lb/in^2, for compactness of the engine and boiler, it is found that complete expansion of the steam in the engine cannot take place in a single stage of expansion. The resulting efficiency is rather low, unless a second stage of expansion be added. Rather than adding a second reciprocating stage, a compact solution would be the use of a rotary positive-displacement stage, of which many forms are possible, e.g. the Wankel type of machine.

180

A useful study of the problems of designing an effective steam car was given by R.M. Palmer to the Institute of Mechanical Engineers on the 10th March 1970. He concludes that a steam car could not compete on technical or operational grounds with a conventional engine, but that, if exhaust emissions are an overriding consideration, a practical steam car does seem to be a possibility in the future. Fuel consumption and acceleration would actually be better than a conventional vehicle at speeds up to 30 miles/h, indicating that, for a city vehicle, steam is a distinct possibility, and currently several designs of steam vehicle, in particular buses, are under development in the USA and elsewhere, using reciprocating, semi-rotary, rotary and turbine expanders.

The major design problems with any vapour cycle include the provision of suffient condensing capacity to cope with continuous full power (about four times that of a comparable IC-engine radiator), the problem of freezing, the control system, lubrication of the expander if not a turbine, and separation of any oil used from the exhaust system. In their favour, vapour cycles can have a limited amount of built-in energy-storage capacity, in the form of thermal capacity in the boiler, sufficient for extra acceleration at starting, while their torque/speed curve is inherently much more favourable than that of an IC engine, since maximum torque is available at starting from rest. These advantages explain the relatively good performance in city conditions, compared with an IC engine.

The steam turbine is the best choice of expander for large powers and low pressures, but, unfortunately, it becomes very inefficient in low powers, so 200 hp or so must be regarded as about the lower size limit, i.e. a bus engine.

This size limitation can be overcome by using a heavy-molecular-weight fluid, e.g. monochlorbenzene (molecular weight = 112·5). This has the effect of greatly increasing turbine efficiency in small powers — the Israelis claim 77% turbine isentropic efficiency at 1 kW — so the prime mover could consist of a directly coupled turbogenerator and feed pump enclosed in a common casing and forming a sealed system with the boiler and condenser. Such a unit offers long life with little maintenance — comparable to the normal refrigerator — and simpler problems of heat exchange than the Stirling engine. No oil is needed for lubrication, since the working fluid itself can be used as a lubricant in hydrodynamic bearings. Being of high speed, the alternator need not be unduly heavy or costly, and, for vehicle use, would be used via a rectifier to charge a battery to provide a hybrid drive.

This turboelectric hybrid drive has received least attention as yet from the manufacturers, but one of the largest Japanese manufacturers is currently considering development.

To sum up: several possible alternatives to the IC engine exist, and could be developed to replace it for particular duties, as follows:

(a) long-range trucks and coaches : gas turbines

(b) high-performance cars : no obvious contender

(c) city buses : steam reciprocating or turbine, Stirling or vapour turbine hybrid

(d) City cars : battery or vapour turbine hybrid.

Since the massive use of large automobiles in the USA is a major factor in the excessive energy consumption of that country compared with the rest of the world, as well as being the chief source of many social evils including smog, ill health through insufficient physical exercise and too much mental stress, and the breakup of the cities with attendant problems of slums and all their problems, and since it is now generally accepted that oil supplies will be scarce and expensive by the end of the century, it is vital that alternatives to the automobile be developed. A viable transport system might include greater use of electric railways (using coal or nuclear energy), clean city buses or commuter cars of the type described and, last but not least, pedal-driven bicycles and possibly tricycles designed for shopping and commuting, with better protection, aerodynamics and provision for luggage than the conventional bicycle.

All this implies a major change in the way of life and in city planning, but gradual change is right and proper, and, if such immense changes in cities have taken place in the last 50 or even 20 years owing to the automobile, equally great changes can be envisaged by the end of the century to stabilise a way of life that is in balance with available energy supplies and better adapted to the proper way of living of the human animal, which has not changed its fundamental nature, mentally or physically, during the 5000 years of so-called civilisation, and, for all practical purposes, never will.

Part 5

Discussion and conclusions, with proposals for a world energy policy

Discussion and conclusions, with proposals for a world energy policy

Prof. M.W. Thring
Department of Mechanical Engineering, Queen Mary College, London, UK.

The object of this study has been to see how the world's limited energy resources can be better used to serve the needs of the whole of humanity now and in the next hundred years.

1   The vital need for a world energy policy

From the two facts (a) that the whole of the developed countries contain only 30% of the world's population and consume 80% of the world's energy and (b) that the fossil fuels of the world are limited, it must immediately be concluded that a few more doublings of the present rate of usage will see the beginning of the resource limitation of these fuels. It follows inevitably that the present free-for-all in the use of the world's energy cannot go on much longer. The rich countries have grown used to the idea that they can use as much energy as they like for any purpose that they can afford financially, while the poor countries can only look on in envy. Such a situation is leading inevitably to a steady increase of world tension, and, eventually, will cause a disastrous situation in the rich countries when they can no longer base their civilisation on an over liberal use of energy (both because of fuel shortage and because of thermal pollution). Thus it is vitally important to prepare a world policy for fuel and energy, taking into account the needs of the less developed countries for a fully adequate standard of living, and their natural desire to have the benefits of technology to an extent fully comparable with the developed countries. In other words, it will certainly be necessary to find a way of giving the real benefits of the first industrial revolution to everyone in the world with a fuel consumption per head considerably below that which is, at the present moment, used in the rich countries – in fact, probably of about the order of that which is the present world average. In this way, the total energy consumption need only double by the end of the century, when the world population will probably be about twice what it is now, but it also means that, in the rich countries, the energy consumption must come down to the present world average figure. This figure is about 1·8 tons of coal equivalent per head per year, while the United States use about 12 tons, and Britain about 5 or 6 tons. The basic problem for countries like Britain and America is, therefore, to provide all the good consequences of the industrial revolution, i.e. freedom to travel and educational benefits, as well as housing and good food and leisure, with an energy consumption of less than a quarter of that which we find necessary at the present time. Enough is better than a feast, because one feasts at the expense of others not having enough.

In this way, it will be possible to extend the oil and coal resources up to the end of the 21st century, by which time we shall have means of using solar energy that can replace them. It is therefore strongly recommended that a world organisation be set up to examine fully all the scientific and technological factors needed to prepare a world energy policy. This organisation must be completely independent of national and industrial pressures, since these tend always to want to use the world's limited energy resources for military or economic reasons that are not in the interests of the world's citizens as a whole. This organisation must be able to present its conclusions based purely on the known scientific factors, specially on a fuller study of all the possible fuel resources, both irreplaceable and replaceable. The Peace Research Institute – SIPRI – is a good precedent for the World Energy Policy Organisation which needs setting up.

Of course, if the world population continues to double every 30 years throughout the 21st century, no policy for energy or food can possibly prevent disaster. Therefore any attempt to introduce a rational and humane control of the world energy usage must necessarily be accompanied by an attempt by all possible humane means to level off the world's population in the 21st century at a figure not greatly exceeding 7,000,000,000.

The rich countries have to change their attitude to energy entirely, from an attitude of spendthrift system based on cheap energy, to regarding energy as something that must be conserved at all costs. Probably they will have to reduce their energy consumption per head to about the present world average of just under 2 tons of coal equivalent per year. At the same time, the poor countries must come up to the same level per capita, since it is impossible to have a world in stable equilibrium divided essentially into the haves and havenots, and certainly not to have a world in which the two-thirds of the world's population are deprived of the necessities for a decent, full, self-fulfilling life.

2   Problems in nuclear-fission power generation

There are a number of unsolved problems in nuclear-fission power generation, which mean that it is not possible at this stage to say that this is the complete answer to our energy-shortage problems, and therefore it is not possible to continue to treat electrical energy as though it was an unlimited cheap resource.

Nuclear fission produces ionising radiation, and radioactive and poisonous materials, all of which have very serious harmful biological effects, particularly on the next generation. Recent research suggests that long-term adverse effects may be caused at much lower levels of dosage than had been previously suspected.

The major unsolved problems for nuclear energy are as follows:

(a)   If we do not use breeder reactors and we are not prepared to excavate vast areas of the Earth's crust for uranium at very low concentrations, the energy obtainable by nuclear fission is limited to a small fraction of that obtainable from coal. I have estimated that it is only about one-hundredth.

(b)   There are still some uncertainties about the safe shutdown of nuclear-fission reactors in the event of a pipe bursting in the light-water reactors, and other less likely failures in the case of the gas-cooled reactors. There is a small, but nevertheless finite, possibility of a release of really substantial quantities of radioactivity into the atmosphere.

(c)   So long as we have the danger of violent revolutions, wars and hijackings, the presence of enriched nuclear fuels in transit from reactors to processing plant and the presence of nuclear reactors in countries with an unstable government, or inadequately trained technical staff, constitute a very real danger of intentional nuclear explosions or a large-scale escape of radioactive materials being produced. Safety in transportation of such materials is in itself a major problem in engineering materials design.

(d) There is at present no known way of storing the nuclear-fission products, which have a high radioactivity for many hundreds of years, that is completely safe. All the present methods require careful looking after by skilled engineers, and it is by no means reasonable or fair to expect our descendants for hundreds of years to look carefully after our waste, which is of no use to them whatever. The highly toxic plutonium inventory of a fast breeder reactor poses a very serious containment problem since it has a half life of 24 000 years and can contain a potentially lethal dose in one piece the size of a tennis ball for the population of the Earth.

(e) There is a small but finite escape of radioactive material from nuclear reactors, and, if the number of nuclear reactors increase sufficiently, this might, in the long run, build up sufficiently to cause local difficulties. It is not known at what level there is no longer any danger from radioactivity, and, in any case, the escaping radio-active materials will be chemically different from those that cause the natural background radiation to which mankind has become accustomed, so that they can more easily enter our bodies via air, food or water.

Since fast-breeder reactors can only intensify the above problems, placing a heavier burden on control systems, the only real remedy open to us at present is to intensify our efforts to solve them before we increase the production of such materials. We should not allow ourselves to fall into the trap of overlooking (or being persuaded to overlook) these very significant problems, in the light of any forthcoming energy crisis (such as the political threat of an embargo on our conventional fuels), in the panic to find an alternative source with which to fill the gap.

3   Liquid and gaseous natural fuels

The British policy in regard to the natural gas that we have found in the North Sea has been to exploit it as fast as possible to recover as quickly as possible the capital invested in drilling for it. This has meant not only that we are converting the whole of the former town's gas system to natural gas, but also that we are using it for every purpose for which it can be economically used. It was even proposed to run power stations from it. In Europe, only about 15% of the crude petroleum is converted into motor fuel, whereas, in the USA, by the use of catalytic crackers, over 50% is converted. Thus the insatiable demand of the US motor car for gasoline has until now been met from the US resources, while the undeveloped countries have not ever had more than a miniscule service of buses. In Europe, we have expanded our use of oil to take over most of the traditional uses of coal — domestic heating, industrial heating, steam raising, gas making, power generation, trains — and have become increasingly dependent on the rich resources of the Middle East for our whole energy system. The only major exception is coke for blast furnaces. In the Middle East, the whole of the natural gas associated with the liquid petroleum has been flared to waste although its energy content is of the same order as that of the liquid.

It is quite certain that we must treat the world's limited resources of oil and natural gas as precious, and make them last as long as possible. Liquid fuel is much the easiest fuel to use for all transport except rail (which can be electrified), and thus, if all our descendants in the 21st century are to have reasonable opportunities for travel and a decent standard of living, we must leave them a reasonable reserve. This requires that the world levels off its rate of petroleum extraction after not more than one more doubling, and that the petroleum is used almost entirely for transport purposes. The world cannot afford competition between different fuels. Private cars with less than four passengers will have to disappear from the scene, and all heavy transport will have to go by rail electrically powered, or by inland waterway.

Natural gas from remote wells will have to be transported and used. Ships full of liquid natural gas constitute such an explosion hazard that the conversion of the gas to a fuel that is liquid at atmospheric temperature and pressure, such as methanol, is probably the best way of using this fuel. Where the natural gas can be piped to an industrial or densely populated urban region, it should be used primarily for the uses for which it is uniquely suited, namely as a chemical raw material, and for uses such as domestic cooking or industrial heating requiring very uniform temperatures, for which its ability to be controlled accurately in very small burners makes it ideal.

Clearly, also, as the liquid and gaseous fuels that flow, or are pumped, from the ground run short, much more research must be done on economical methods of using the tar sands of Athabasca and the oil shales of Colorado or elsewhere as sources of liquid fuels, petrocoke and gaseous fuels.

4 Solid fuel

Since the world's resources of solid fuel in seams of adequate thickness to be worked at least by machines are considerably more than ten times the liquid and gaseous fuel resources put together, and since these resources are much more widely spread over the Earth's surface, it is a necessary part of a long-term world fuel policy to find a satisfactory way of using these resources.

The main problems of using coal are:

(a) The fact that a mine operated by humans is immensely more expensive than an oil well, and, even when all precautions have been taken, there are still many lives lost during a life spent getting to and from the face, and working at it is not a decent life for humans – in Roman times, this was the one activity that caused slaves to risk the death penalty by running away.

(b) Pollution from the products of combustion: If the world's combustion of hydrocarbon fuels were to double two or three times more, the CO_2 content of the atmosphere would rise to a level at which it would have significant effects on the environment, such as the temperature rise due to the greenhouse effect. The much more immediate effects are those due to the products of incomplete combustion (which can in every case be cured by better fuel technology, nearly always with a reduction in energy consumption) and those due to oxides of sulphur.

The mining problem can and must be solved before the end of this century by the development of mole miners, underground gasification, or underground combustion to produce electricity. The SO_2 emissions must be eliminated by pretreatment of the coal to remove S or by dry or wet-cleaning the combustion gases with, for example, dolomite.

5 Fuel economy

Many human purposes, such as domestic heating or transport, can be served as well, or very nearly as well, with less than half the energy consumption that we use in a period of regarding energy as an unlimited cheap resource. Industrial heating and steam usage can be reduced by at least 25%. During the Second World War, in Britain fuel economy committees achieved this without causing serious hardship. If, for example, as proposed by Sir Oliver Lyle at that time, a tax is put on all primary fuels that doubles or triples their cost, it at once becomes economic to install fuel-saving devices (e.g. air heaters, economisers, higher-efficiency boilers, furnaces or stoves, better insulation), and people would stop wasting fuel for trivial purposes; they could walk or cycle to the shops, and switch off unnecessary lights.

The use of pass-out steam from power stations for industrial purposes and domestic heating lowers the efficiency of electricity generation, but produces very large overall fuel savings, specially when the electricity has previously been used to heat the houses. It also reduces the size of the cooling towers needed, and eliminates the enormous evaporation of fresh water associated with these towers.

Industrial production can be reduced to use about one-quarter of the energy at present used per head simply by making consumer goods that last much longer and are designed to minimise fuel consumption in manufacture, and by realising that the world cannot afford the energy for gimmicks and goods people are persuaded to buy, to keep up with the Joneses. More recycling of waste materials and reuse of containers can also save a great deal of fuel. Perfectly adequate travel and transport of all necessary goods can be provided with considerably less than one-quarter of the present rich-country consumption per head and with no air pollution by P_b, C_nH_m, or CO, by the use of existing knowledge for small engines, buses, electrical trains, bicycles and hybrid diesel-electric vehicles. Further vehicle-engine research aimed at fuel economy instead of acceleration could improve this figure still more. Canals will be used again, since water transport uses the least energy of any method for heavy goods. Pipelines also provide a very low energy method of transporting liquids and gases, and may also be used for solids, such as coal suspended in water or oil.

Domestic heating requirements can be enormously reduced by better insulation and draft control; solar heating of water with fuel topping up can reduce the fuel consumption very much. A school in the English Midlands has been successfully heated entirely by solar energy, and this principle can save a good part of the heat, provided summer overheating can be avoided by good control. Where electricity must be used for domestic heating, the heat pump can be further developed to give an apparent first law efficiency of 300% or more.

Present agricultural practices in rich countries are equally based on a lavish use of fuel for tractor and other machine power, for crop drying and for the fixation of nitrogen. We are also using up the readily available rich sources of P and K, so that a long-term equilibrium growth of food will require entirely different methods based on much more recycling of organic matter, less runoff into rivers, use of solar energy for drying, and much-more-low-power animal or hand-operated cultivation and cropping. The production of methane by anaerobic decomposition of biological refuse gives a fuel byproduct during a stage of conversion of this material into natural fertiliser or compost. This process will have to be developed and applied to all such material from cities.

6   Research priorities

In addition to the two major lines of research mentioned above (making nuclear fission in breeder reactors completely safe and finding a way of bringing solid coal to the surface without men going underground), the following additional lines of research seem to need much more attention than they are receiving:

Development of

(a) small power-energy units for places remote from large power stations, using wind power, local combustibles such as wood refuse, solar power, water power, and methane.

(b) large area solar-power units of moderate capital cost per kilowatt.

(c) large area solar stills, so that deserts can be irrigated from sea water, and crops with a fuel residue such as sugarcane can be grown.

(d) energy storage by electrolysis of water, a suitable chemical storage system for H_2, and a fuel cell for reconversion to electricity.

7   Conclusion

It seems clear that, unless a rational world fuel and energy policy broadly along the lines laid down above is agreed in the next 15 or 20 years, the rich countries will have done irreparable harm to the possibility of mankind coming into stable equilibrium with the environment and living a life of decent quality in the 21st century and after.

Energy contents, units and equivalents

Energy contents, units and equivalents

In the preceding papers, the authors have expressed quantities in terms of different units, either from personal preference or because of accepted conventions. To allow comparisons to be made, the following tables have been included to relate the various energy sources and units.

Table 1

Average energy content of the different fuel forms

Coal	12500	Btu lb^{-1}
Oil	19000	Btu lb^{-1}
Natural gas	1000	Btu scf^{-1}
U^{238} Fission	8·1 x 10^{10}	J g^{-1}
(Breeder)	35 x 10^{9}	Btu lb^{-1}

Table 2

Coal equivalents, tce (UK tone coal equivalent)

1 g U^{238}	= 2·7 tce
1·0 toe	= 1·5 tce
28 x 10^{3} scf NG	= 1·0 tce
(8·2 x 10^{3} kWh (heat)	
(28 x 10^{6} Btu	= 1·0 tce

Table 3

Useful conversion factors

1 tonne	= 1000 kg
1 UK ton	= 1016 kg
1 US ton	= 907 kg
1 ton oil	= 7.5 barrels
1 barrel	= 35 Imperial gallon
	42 US gallon
(1 Imperial gallon	= 1·2 US gallon)
1 tone crude oil	= 260 gallon
1 cal g^{-1}	= ·1·8 Btu lb^{-1}
1 mile	= 1·609 km
1 hectare	= 10 000 m^2
	2·471 acre

Table 4

Energy equivalents

1 joule (J)	$\equiv$	1 Newton meter (Nm) $\equiv$ 1 watt second $\equiv 10^7$ ergs
1 kilocalorie (k cal)	$\equiv$	10^3 calories $\equiv 4186 \cdot 8$ joules
1 Btu	$\equiv$	$251 \cdot 996$ calories $\equiv 1055 \cdot 06$ joules
1 therm	$\equiv$	10^5 Btu $\equiv 29 \cdot 307$ kilowatt hour (kWh)
1 kilowatt hour	$\equiv$	$3412 \cdot 14$ Btu $\equiv 1 \cdot 341$ horsepower hour (hph)
	$\equiv$	$859 \cdot 845$ kilocalories $\equiv 3600$ kilojoules
1 ton coal equivalent	$\equiv$	$2 \cdot 8 \times 10^7$ Btu $\equiv 8 \cdot 2 \times 10^3$ kWh
(tce)	$\equiv$	7 tons TNT

References

1.   SPIERS, H.M.: Technical data on fuel, 6th edn. British National Committee, World Power Conference, 1962

2.   'Efficient use of fuel' (HMSO, 1958, 2nd edn.)

3.   KING HUBBERT, M.: 'Energy resources' in 'Resources and man'. Committee on Resources and Man, National Academy of Sciences, National Research Council (W.H. Freeman, San Fransisco, 1969) p.157

Index

9/ A6